AP* Achiever
Advanced Placement Environmental Science Exam Preparation

to accompany

Environmental Science: A Global Concern

Ninth Edition

Bill Cunningham
Mary Ann Cunningham
Barbara Woodworth Saigo

Prepared by
Margaret Scot Smith
Berkeley Preparatory School

Mc Graw Hill Higher Education

Boston   Burr Ridge, IL   Dubuque, IA   New York   San Francisco   St. Louis
Bangkok   Bogotá   Caracas   Kuala Lumpur   Lisbon   London   Madrid   Mexico City
Milan   Montreal   New Delhi   Santiago   Seoul   Singapore   Sydney   Taipei   Toronto

The McGraw·Hill Companies

Advanced Placement Environmental Science Exam Preparation to accompany
ENVIRONMENTAL SCIENCE: A GLOBAL CONCERN, NINTH EDITION
BILL CUNNINGHAM, MARY ANN CUNNINGHAM, AND BARBARA WOODWORTH SAIGO

Published by McGraw-Hill Higher Education, an imprint of The McGraw-Hill Companies, Inc., 1221 Avenue of the Americas,
New York, NY 10020. Copyright © 2007 by The McGraw-Hill Companies, Inc. All rights reserved.

1 2 3 4 5 6 7 8 9 0 QPD/QPD 0 9 8 7 6

ISBN-13: 978-0-07-325689-4
MHID-10: 0-07-325689-7

www.mhhe.com

* Pre-AP, AP and Advanced Placement program are registered trademarks of the College Entrance Examination Board, which was
not involved in the production of and does not endorse these products.

Table of Contents

Appendix
A: Important Environmental Legislation
B: Major Pollutants Chart

SECTION III: WRITING ESSAYS

Student Tips for Writing AP Environmental Science Essays

SECTION IV: PRACTICE TESTS

Practice Test I
Practice Test I Answers
Practice Test II
Practice Test II Answers

Margaret Scot Smith (Scottie), Berkeley Preparatory School

Scottie Smith teaches AP Environmental Science and Anatomy and Physiology at Berkeley Preparatory School in Tampa, Florida. She received a BS in Microbiology from Auburn University and a PhD in Microbiology and Infectious Disease from Texas A&M University. She has been an AP ES Reader since 1998, a Table Leader since 2002, and is also a College Board Consultant. She has taught environmental science courses in high school and college for 15 years. She is very involved in the Envirothon, North America's largest environmental competition, and has led excursions to Ecuador, the Galapagos Islands, and the Florida Keys.

Introduction to Advanced Placement Environmental Science

The Advanced Placement (AP) Environmental Science course will be very demanding of your time, as it is a college-level course for which you might receive college credit. The class is designed to be a one semester, entry-level environmental science course with a laboratory component. The class will address historical, ecological, social, political, and economical ramifications of the global environmental crisis. The course has a rather broad scope, including biology; chemistry; physics; earth science; and the social sciences including history, economics, and politics. You must have an excellent work ethic, as the class is very thorough and potentially difficult due to its varying topics. The ultimate purpose of the class is to take the AP exam next May and score a grade acceptable to gain college credit. The goal is for you to learn about your environment and make you aware of our current global situation.

Although the course crosses many disciplines, there are six themes that reinforce the material learned in the class. First, the scientific method is an intrinsic part of any science course. The exam will require you to be able to apply and explain the parts of the scientific method. The next unifying thread is explained in chapter 3 of this review guide—the first and second laws of thermodynamics. Energy cannot be created nor destroyed, and all energy proceeds toward entropy. Energy conversions are essential to maintain life. The idea that the earth is an allied system that changes and allows matter cycling is also a common theme. The role of humans in the environment is another common thread throughout the course. Although humans have been on earth for hundreds of thousands of years, only recently has technology allowed us to increase our population to the point where our ecological footprint is far larger than in any point in earth's history. The historical, social, political, and economical issues that help create our environmental problems will also be the resources that we use to solve those problems. This thread is linked to the final thread of sustainability. To allow humans to survive on earth we must learn to engage in sustainable development and conservation science while avoiding damage to the earth's common resources. These unifying themes will be underlying all material studied in conjunction with this course.

The exam is written by a committee of college professors and high school teachers. They work together to generate an exam that will have 100 multiple-choice questions and four essays. The exam takes three hours—one and a half hours each for the multiple choice and free-response questions.

Sixty percent of the student's score will come from the 100 multiple-choice questions chosen by the development committee to address the topics in the AP Environmental Science course description published by College Board's Advanced Placement program. The topic outline in the course description lists the percentages that will indicate the number of questions that can be expected to compose the 100 multiple-choice questions. The multiple-choice questions range from quite simple evaluation of data sets to rather difficult synthesis and analysis questions. The multiple-choice portion of the exam is scored by a machine.

To date, all AP Environmental exams have begun with a short section I think of as intrinsic matching. You are given five choices and four to five questions about the choices. The answers may be used once, more than once, or not at all in each of these sections. The 20 or so question matching section is then followed by approximately 80 multiple-choice questions. When taking a multiple-choice test, you should not guess an answer if you have no idea of the answer. Conversely, if you can narrow down your choices to two to three possible selections, choose what you think is the best answer. As on the SAT, you will lose one-fourth of a point for each answer you get wrong, but you get a full point for each question you answer correctly. Therefore, if you can narrow down your choices, it would be best to answer the question.

The four free-response questions will count for the remaining 40 percent of the student's grade. The test will have one document-based question, two synthesis and analysis questions, and one data-set question. AP Environmental questions are not one question that can be answered with a five-paragraph essay. They have numerous sub-questions within them that must be addressed to garner full points. The document-based question will be a short paragraph, usually from a hypothetical newspaper, that the student must read and use to answer a series of pointed questions. The questions will not only be about the article, but on issues peripheral to the contents of the article. The data set question may be a graphing question where students are required to graph data, or it may be a question where students are required to interpret graphed data. The question could involve numbers for which students are asked to make calculations and derive solutions. Students must also address the additional questions brought about by an interpretation of the data. The other two questions are synthesis and analysis questions. These questions usually incorporate a short paragraph, or perhaps a diagram, that the students must use to answer the subsequent questions.

High school and college faculty from across the United States and Canada meet in one location each summer to grade the free-response questions. These people are called readers and the head of the group is known as the chief reader. Some highly experienced readers function as the leaders of the readers and are called question-and-table leaders. The question-and-table leaders arrive early to work together to create a rubric, an answer key with most possible answers that students might come up with in answer to a question. The rubric has nearly every conceivable answer that students might put in response to a question. The rubrics are multi-page detailed responses to the questions that incorporate even regional answers if a question might be answered differently in different geographic areas of the United States. Question-and-table leaders derive their rubric based not only upon what they expect from student answers but by reading numerous student responses to determine how students interpreted the question. Once the readers arrive, they are taught how to apply the rubric and spend the next seven days grading all the AP Environmental exams taken that year. No changes are made in the rubric once the grading has begun so that all exams are graded fairly. Grading the exams all at once ensures consistency in grading. Students are evaluated equally based upon these published rubrics.

Once the multiple-choice portion has been graded and the essays scored, the College Board converts the added raw scores into composite scores from one to five with five being the best possible score.

This review guide is designed to help you in preparing for the AP exam in May. It consists of chapter reviews followed by a series of multiple-choice questions with detailed explanations of the answers. The chapter reviews list important skills required to understand each chapter, as well as the vocabulary necessary for the section. The essay section gives a detailed explanation on how to write AP Environmental essays and practice writing essays and reading rubrics. There is a reference section on common pollutants and a chronological listing of important environmental legislation. Finally, there are two three-hour practice exams that will allow you to practice your timing prior to the May test.

The review guide is based upon the topics from the course description for AP Environmental Science as presented by the College Board. As a result, although the chapters in textbook *Environmental Science: A Global Concern* have equivalent numbers of pages, the study guide chapters vary in length. For example because population ecology makes up from 10-15 percent of the exam, those review chapters 6 and 7 are quite long and contain more detail than the introductory material found in chapter 1 of the review guide.

The "Take Note" sections of this review guide are designed to help you prepare for the AP exam. They may suggest focusing on a particular topic or ensuring that certain material is not misunderstood. They will frequently mention questions that have appeared on previous AP Environmental exams. These sections are not intended to be your main focus as you prepare for this exam; they are simply to call attention to certain aspects of the material.

Chapter 1 - Understanding Our Environment

Key Terms

biocentric preservation	human development index
ecological footprint	indigenous people
environment	sustainable development
environmental science	utilitarian conservation

Skills

1. Distinguish between the environment, environmental science, and ecology.
2. Summarize the contributions of various individuals in the environmental history of the United States.
3. Identify the current global environmental problems that humans are facing.
4. Argue whether sustainable development is possible. Evaluate the benefit of economic growth over environmental damage.
5. Examine the plight of indigenous people. Appraise their value in terms of earth's diversity.

Introduction

The environment is an organism's surroundings. Environmental science is not just the study of ecology, the interaction of living things in their environment. Environmental science also addresses the role of human beings and the effects their activities have on the environment.

Brief History of the Environmental Movement in the United States

- George P. Marsh
  - published *Man and Nature* in 1864, which warned of environmental damage occurring in the West
  - partially responsible for establishment of the forest reserves in 1873
- Theodore Roosevelt
  - noted for his interest in conservation
  - established the first wildlife refuge to protect the brown pelican
  - moved the Forest Service into the Department of Agriculture from the Department of the Interior
- Gifford Pinchot
  - first head of the forestry service under Roosevelt
  - used utilitarian conservation—use the forests for the greatest good for the greatest number of people for the longest possible time—still used by the National Forest System (NFS) as the multiple use sustained yield policy
- John Muir
  - founded the Sierra Club
  - believed in biocentric preservation—living things have intrinsic value and therefore ought to be protected simply because they exist
  - assisted in the establishment of Yosemite and Kings Canyon National Parks

- Stephen Mather
  - follower of John Muir
  - first head of the National Park Service established in 1916
- Rachel Carson
  - biologist who published the seminal *Silent Spring* in 1962, which suggested that the levels of DDT were causing the declines of raptors
  - initiated concern into the effects of pollutants on the environment

> **Take Note:** The history of the environmental movement is important. Every chapter will have more historical background and pertinent legislation that must be understood to do well on the AP exam in May.

Current environmental concerns

The first and foremost concern is the alarming increase in the human population. All other environmental problems in the world can be attributed to the distribution and density of humans in the natural environment. The current human population is over 6 billion and is expected to climb to 10 billion in the next 50 years. Critical problems for this looming increase in population include the following:

- Food production
  - soil erosion—movement of soil from one area to another
  - salinization of farmland—increasing irrigation results in deposition of salt on the surface of agricultural lands
  - pesticides and fertilizers contaminate ground and surface water
- Decrease in biodiversity of habitats
  - species extinction—the irreversible loss of a species when the last of its members die
  - deforestation—removal of forests for agriculture, grazing, and lumber/wood
  - destruction of wetlands
  - desertification—degradation of previously viable land by overuse due to agriculture or ranching
  - invasive exotic species—non-native species that lack competition or predators in a new environment so that they rapidly proliferate
- Poverty
  - over one-sixth of the earth's human population live in extreme poverty and lack sanitation, drinking water, adequate food, healthcare, education, energy resources and housing
  - poor are frequently subject to environmental injustice as they do not have the education nor resources to resist contamination or degradation of their environment
  - 20% of the earth's population lives in the 20 wealthiest countries
  - human development index—evaluates an individual's quality of life based upon education, sanitation, survivorship, and income
  - wealth gap—the gap between the wealthy and the poor—has been steadily increasing

- Energy
  - depletion of fossil fuel reserves
  - pollution from combustion of fossil fuels
  - environmental damage from recovery/transport of fossil fuels
- Water
  - groundwater/surface water contamination
  - groundwater depletion from pumping for irrigation
  - surface-water shortage
- Pollution
  - global warming—an increase in the earth's temperature due to increased levels of greenhouse gases, which retain infrared radiation, thus heating the atmosphere
  - acid deposition—deposition of low pH precipitation resulting from acidic particles released into the atmosphere
  - ozone depletion—loss of the protective stratospheric ozone due to the release of chlorine containing compounds into the atmosphere

> **Take Note:** This introductory overview is cursory. Far more detail regarding these pressing issues will be addressed in upcoming chapters.

Sustainability

Sustainability means using the earth's resources in a responsible fashion, thus allowing current and future generations access to the resources. Sustainable development means improving one's current standard of living without compromising the standard of living of future generations. Economic growth must be balanced against environmental degradation. The ecological footprint is the measure of the demands people place upon nature. A questionnaire allows you to determine the impact you are making upon the earth. Global environmentalism is examining global environmental issues as a whole, not as separate problems in order to work toward solutions in conjunction with other countries.

Indigenous People

Indigenous people are people native to an area that tend to be subject to environmental injustice and repression. These individuals are frequently oppressed by newcomers and then lose their land, food resources, and culture.

Chapter 1 Questions

Match the individual to his contribution to the environmental movement.
   a. John Muir
   b. Gifford Pinchot
   c. Theodore Roosevelt
   d. Stephen Mather
   e. George Marsh

1. ran U.S. Forestry service under concept of utilitarian conservation
2. believed in preservation of natural areas for the benefit of future generations

3. first director of the National Park Service
4. established the first wildlife refuges in the United States

5. Which of the following is an example of an invasive exotic species?
   a. Mule deer in the southwestern United States devour their food resources and begin to die out.
   b. Salmon stock in the Pacific northwest have difficulty traveling upstream due to dam placement along the Columbia River.
   c. Giant pandas in China are consuming bamboo at an alarming rate in a small area set aside for preservation.
   d. South American pampas grass in Southern California disperses seeds from its long plumes over long ranges.
   e. Bald cypress trees in south Florida resist desiccation because they shed their needles during the winter dry months.

6. Which of the following describes the greenhouse effect?
   a. Heat from space enters the earth's atmosphere, thus warming the earth.
   b. Infrared heat from the earth's surface cannot escape the atmosphere of the earth and warms the planet.
   c. The sun's rays heat the atmosphere, causing the greenhouse effect.
   d. The reflection of sunlight from the water and ice on the earth's surface causes the earth to have a warm atmosphere.
   e. The earth's atmosphere creates an impenetrable barrier to the heat generated by radioactive isotopes in the earth's crust.

7. The Kyoto Protocol has as its goal
   a. decreasing CFC use by all countries by 2020.
   b. reducing greenhouse gases.
   c. ensuring safe drinking water in developing countries.
   d. increasing global food production.
   e. preventing the use of DDT as a pesticide.

8. The leading cause of species endangerment is
   a. maintenance of native species in an area.
   b. fishing limits placed upon fish found in low number in the ocean.
   c. habitat destruction.
   d. preventing the harvest of old growth forests.
   e. preventing environmental contamination with DDT.

9. Fossil fuels provide what proportion of the energy used in industrialized countries?
   a. 40%   b. 50%   c. 60%   d. 70%   e. 80%

10. The poorest nations are characterized by which quality of life indicator?
    a. low infant mortality rates   b. low income   c. low number of children
    d. adequate nutrition   e. high availability of medical care

Explanations

1. b
2. a
3. d
4. c
5. d. Invasive exotic species are those that are not native to an area. Mule deer are native to the U.S. southwest, salmon are native to the Pacific northwest, giant pandas are native to China, and bald cypress are native to the southern United States. The only exotic mentioned is the South American pampas grass growing in California.
6. b. Space is cold, not hot. The sun's radiation is absorbed by the surface of the planet and serves to warm the planet. As the heat builds, it cannot escape back out of the atmosphere into space due to the high level of gases that prevent its diffusion into space. The reflection of sunlight would cool the planet. The sun's energy heats the earth; it is not warmed internally.
7. b. The Kyoto Protocol has for its purpose lowering atmospheric greenhouse gases.
8. c. Habitat destruction is the leading cause of species endangerment. Maintaining native species in an area should help prevent endangerment. Placing fishing limits on fish found in low number in the ocean will help avoid their endangerment. Preventing the harvest of old growth forests will protect habitat, and preventing environmental contamination with DDT will prevent bioaccumulation in sensitive species.
9. e. Fossil fuels provide 80 percent of the energy used in industrialized countries.
10. b. Low income is the quality of life indicator characterized by the poorest nations. The quality of life indicators found in the poorest nations include high infant mortality rates, low income, high numbers of children, inadequate nutrition, and low availability of medical care.

Chapter 2 - Frameworks for Understanding: Science, Systems, and Ethics

Key Terms

anthropocentric	equilibrium	negative feedback loop
biocentric	ethics	positive feedback loop
blind experiment	field investigation	preservationist
controlled experiment	frontier science	resilience
deductive reasoning	hypothesis	scientific consensus
dependent variable	independent variable	stewardship
disturbances	inductive reasoning	systems
double blind experiment	instrumental value	theory
environmental justice	intrinsic value	toxic colonialism
environmental racism	models	utilitarian

Skills

1. Diagram the stages required to test a hypothesis.
2. Identify independent and dependent variables in a controlled experiment.
3. Conclude what criteria are necessary for a hypothesis to become a scientific theory.
4. Contrast deductive and inductive reasoning. Give an example of each.
5. Compare and contrast positive and negative feedback loops. List specific environmental examples of each type of loop.
6. Characterize the different philosophical viewpoints of anthropocentric, biocentric, utilitarian, and preservationist worldviews.
7. Examine the concepts of environmental justice and environmental racism.

The Nature of Science

Science is a method of producing knowledge in a systematic fashion based upon observation of natural phenomena. Scientists conduct experiments or field observations to explain their proposed explanations. Science qualifies all results of experiments as conditional, because there is always the outside chance that additional research will demonstrate the results to be inaccurate. Scientists never "prove" anything, they "demonstrate," "show," "correlate," and "illustrate." For a scientist to publish his results, he must submit his research data to a peer-reviewed journal. Scientists must also do each experiment multiple times, known as replication. His or her work must be able to be repeated by other researchers, called reproducibility. For example, in 1989 two scientists claimed at a press conference to have carried out cold fusion in the laboratory. The energy and scientific communities were skeptical and tried to reproduce the data. They could not, nor was the experiment ever published in a peer-reviewed journal. Due to the lack of reproducible data, cold fusion was not determined to be possible at that time.

Reasoning

There are two types of reasoning, deductive and inductive. Deductive reasoning proceeds from the general to the specific. For example, if you know that venomous snakes typically have a triangular head and while hiking you find a snake with a triangular head in your path, deductive reasoning tells you that you should avoid the snake as it is likely venomous. Inductive reasoning is making a general rule from numerous observations. For example, it is well known that poison dart frogs are vibrantly colored. Other toxic species include the venomous sea slugs, the noxious tasting ladybug, and many species of caterpillars, all three brightly colored. Therefore, using inductive reasoning, brightly colored species tend to be toxic. This coloration is known as warning or aposematic coloration because the brightly colored animals are usually toxic.

Designing an Experiment

> **Take Note:** A "design an experiment" question is to be expected on each AP exam. You will be given a scenario and must develop a controlled experiment to test a hypothesis. You absolutely must be able to describe your hypothesis and experimental setup, including constants, controls, data, and interpretation, to answer such questions properly. See further information in the section on writing essays.

There are a series of steps in experimental design.

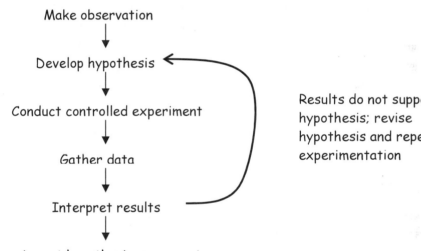

Make observation

↓

Develop hypothesis

↓

Conduct controlled experiment

↓

Gather data

↓

Interpret results

Results do not support hypothesis; revise hypothesis and repeat experimentation

↓

Accept hypothesis as correct

Once an explanation has been supported throughout the years with numerous studies, the explanation is deemed a scientific theory. This use of the word theory is as close to absolute as a scientist will come to stating something as fact. People not familiar with science tend to use the word theory as conjecture, but it means the opposite to a scientist. Theories include the theory of evolution and the theory of relativity, both substantiated by large amounts of scientific evidence and accepted by the scientific community. Theories may be modified as additional data is collected.

When conducting experiments, you can conduct statistical analysis from available data. For example, you can count how many West Indian manatees are found dead in a given year and calculate the statistical relevance of the number by comparing it to the number found dead in prior years. This procedure is known as a natural experiment, because it analyzes previous observations.

Most frequently scientists must manipulate an experiment to derive the data in which they are interested. In that event, they must use a controlled, or manipulative, experiment, in which only one factor is changed and all the rest of the factors are constants. The factor that is changed is usually known as the variable. All other parameters of the experiment remain constant. This setup is known as a controlled experiment. For example, an observation is made that a species of frog is declining in an unpolluted area. Researchers cultured an unusual fungus from the bodies of the dead and dying frogs. To test whether the fungus was killing the frogs, a controlled experiment must be conducted. Two tanks with the same number of frogs of the same species would be set up. The tanks would have identical temperatures, water pH, water volume, light, and oxygen. The animals would be fed the same amount of food. In one of the tanks the fungus would be introduced. If the animals died, then the fungus could be killing the frogs, and the hypothesis would be accepted. The experimental or treatment group is the group exposed to the fungus. The control group of frogs would be the group not exposed to the fungus. The independent variable is the factor being changed and the dependent variable is the result that occurs in response to a change in the independent variable. The independent variable in this example is the fungus and the dependent variable is the death of the frogs.

To alleviate researcher bias, the experiment could be conducted as a blind study, where the researcher did not know which tank contained the fungus. Frequently a double blind study is conducted when testing new vaccines or medications. Neither the researcher nor the study participant knows if they have been given the medication. Then, if in the control group four people got headaches and in the experimental group three people got headaches, headaches are not likely to be a side effect of the medication. If the researcher or study participant knew which treatment they received, it might cause them to imagine or perceive a headache that did not exist.

Probability and Statistics

Probability is the likelihood that something will happen. The probability of two events occurring simultaneously is the product of their individual probabilities. For example, the likelihood of a tossed coin coming up head is one-half, because there are only heads and tails. The probability that two coins tossed together would both come up heads is one-fourth, or one-half x one-half. Statistics focus on calculating probabilities that the results of an experiment are accurate and do not arise as a result of chance.

Models

Models are representations. For example, a graph is a graphical model that illustrates relationships. Physical models are a miniature representation of an object. For example a model car or a model of the solar system would be physical models. The most accurate models are mathematical, because they are always true. For example, $E=mc^2$ is always an accurate representation of the mathematical relationship between energy, mass, and the velocity of light.

Consensus vs. Frontier Science

Frontier science is relatively new research that has not yet been supported by years of scientific evidence. For example, nearly ten years ago an article was published in a medical journal that established a link between autism and childhood immunization. The article has since been repudiated, as there has been no research that substantiated the researchers' claim of a link. The more current research seems to demonstrate that it is coincidence that autism appeared after immunization, rather than a cause-and-effect relationship. Unfortunately, many people with newborn children elected not to have their children vaccinated, which has lead to an increase in easily prevented diseases such as measles and whooping cough. Consensus science is science supported by large amounts of data. For example, most scientists agree that the increasing levels of CO_2 in the earth's atmosphere are causing more heat to be retained in our atmosphere, resulting in global warming. Paradigm shifts are large changes in explanations as science uncovers more information. Plate tectonics not only greatly changed the science of geology, it also had a profound impact on paleontology and evolutionary biology.

Systems

A set of interacting components forms a system. For example, your stereo system has a receiver, CD player, and speakers. An ecological example of a system is a pond. The plants and algae present provide food for the numerous fish. The wastes and decay from the fish and plants allow the nutrients to cycle in the system. If the system changes very little, it is said to be in equilibrium. A disturbance is an alteration in the equilibrium. For example, a drought might dry up the pond a bit. Periodic mild disturbances may help an ecosystem remain at equilibrium. If a system can recover quickly, it is said to be resilient. An open system receives input from the outside, but a closed system is not open to the outside for input. The earth is an open system for energy because it receives energy from the sun, but the universe or an isolated atom is a closed system.

Feedback Loops

Take Note: Students frequently confuse positive and negative feedback. One way to think of it is that positive is bad and negative is good. Unfortunately this thinking is contrary to the way you would normally envision those terms. Be sure you have a thorough understanding of these mechanisms, because feedback loops maintain homeostasis in systems.

Feedback loops are found in systems that allow for correction of that system. A negative feedback loop is one in which the deviations from a normal set point are minimized. A good example is predator prey relationships. When the number of prey increases, the number of predators increases after a slight delay. When the predator increase causes the prey number to drop, the predators die out, thus maintaining both populations around a standard set point. Negative feedback loops help maintain homeostasis, or the normal steady operating state of an organism or an ecosystem. A positive feedback loop is one in which the deviation from normal becomes greater and greater instead of correcting itself.   In organisms, disease may result from positive feedback. A good ecological example of a positive feedback loop is eutrophication (eu—true; troph—feed) in a pond. In eutrophication, an increase in nutrients induces an algal bloom. The bloom blocks the sunlight from rooted vegetation and deeper algae, resulting in death. The subsequent decay of the dead plants and algae consumes the dissolved oxygen in the water, causing species with higher oxygen requirements to suffocate. Those dead organisms decay, further reducing the oxygen. The anoxic (an—without; oxy—oxygen) conditions eventually result in all living organisms dying out in the pond.

Ethics and Worldviews

Ethics is the branch of philosophy that deals with right and wrong. Environmental ethics is examining our principles that guide us to do what is right in our actions toward the environment. Stewardship is managing and caring for the environment. A biocentric (bio—life; centr—center) worldview is that living things have a right to exist and should be respected. An anthropocentric (anthro—man; centr—center) worldview is that humans are above nature, not a part of nature. A utilitarian worldview believes that nature provides services and goods to humans for our use. Conservationists believe we should care for our environment so that it may provide for us for many years to come.   Gifford Pinchot, the first head of the United States Forest Service (USFS), was appointed under Theodore Roosevelt because he was a noted conservationist. Preservationists feel that the environment should be protected in its natural state, but not used for the good of humans. John Muir, founder of the Sierra Club, was a preservationist.

Many people feel that because a living organism exists, it has value. This concept is called intrinsic or inherent value. If something has value because you can use it for a purpose, it has instrumental value. If something is considered to have value because it is beautiful, it is said to have aesthetic value.

Ecofeminism is the concept that males are more domineering and try to subjugate nature, while women are more nurturing and protective. Therefore, if women were not repressed in societies, nature would not be as exploited because women would have more of a voice in environmental issues.

Environmental justice is giving people of all socioeconomic, racial, ethnic, and religious affiliations equal rights with regard to environmental protection. For example, there are greater problems with lead poisoning in lower socioeconomic and ethnically diverse populations than in upper-income Caucasian populations. Environmental racism is the greater exposure to environmental pollutants based on race. Examples include the farm workers of color working on U.S. farms and being exposed to high levels of fertilizers and pesticides or dumping hazardous wastes on Native American reservations. Toxic colonialism is shipping toxins to reservations or undeveloped countries that welcome the income and disregard the damage that may result from the toxins. The U.S. Office of Environmental Justice was established by the Environmental Protection Agency (EPA) in 1992 to assist groups with fighting emissions or toxins in their area when the groups do not have the resources to fight by themselves. It helps identify at-risk areas and the specific risks to the area.

Chapter 2 Questions

Use the paragraph below for questions 1 and 2.

A scientist wished to test the impact of increased nitrates on plant growth. He placed 100 rye seeds in three different pots, making sure to use the same soil in each pot. He placed 100 ml of nitrate free fertilizer in pots 1 and 2 and nitrate containing fertilizer in pot 3. He also added clover seeds to pot 2 (clover is a legume). He placed the plants under a grow light and watered them daily with 50 ml of water for three weeks. His results are shown below.

Pot	Average ht rye (cm)	Increase in mass (g)
1	4.2	100.4
2	6.8	163.0
3	7.6	172.2

1. What is the independent variable in this experiment?
a. soil   b. height   c. mass   d. nitrates   e. rye seeds

2. What statement best explains the use of pot 2 in the experiment?
a. The nitrogen-free media only causes increased growth of the plants if other seeds are present.
b. Clover seeds inhibit the growth of the rye grass.
c. The clover must be fixing nitrogen from the atmosphere and enriching the soil thus enhancing the growth of the rye grass.
d. The clover seeds crowd the rye grass and force it to grow better in their presence.
e. The nitrogen-free media is toxic to the rye seeds but not the clover seeds.

3. Utilitarian conservationists believe in protecting resources for their
a. value as habitat for wildlife.   b. aesthetic value.
c. intrinsic value.   d. economic value.
e. role in the biosphere.

4. Which of the following is an example of environmental racism?
a. Sewage sludge from a sewage treatment plant is incinerated to generate electricity.
b. Numerous hazardous waste sites are located on land owned by Native American tribes.
c. Pesticide residues may be found on some fruits and vegetables.
d. Lead can be removed from soil by using plants such as sunflowers that will take up the lead in their tissues.
e. Dredging channels in rivers removes not only sediment but varying amounts of polychlorinated biphenyls and remnants of oil spills.

5. A hypothesis is
a. a testable explanation for an observation.
b. a theory devised after years of scientific study.
c. proven true by general agreement of the scientific community.
d. always demonstrated to be accurate after one experiment is conducted.
e. rejected when the data support the premise.

- 12 -

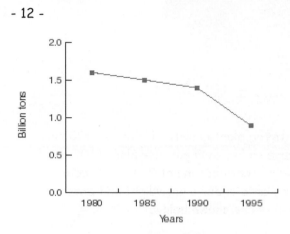

(http://www.in.gov/idem/soe/99report/air/Air09.html)

6. The above graph shows U.S. sulfur dioxide emissions over a 15-year period. What can be inferred from the graph?
a. The enforcement of the Clean Air Act passed in 1990 resulted in less sulfur dioxide emissions than in the previous decade.
b. There was a 50 percent decline in sulfur dioxide emissions from 1980 to 1990.
c. It can be determined from the graph trend that only 0.5 billion tons were released in 2000.
d. Most of the sulfur dioxide was from natural rather than anthropogenic sources.
e. Sulfur dioxide emissions are linked to increased global warming.

7. A scientist conducted a study to determine which wildflowers in an area attracted honey bees. She conducted visual observations in four different fields and recorded the number of visits by bees to each of the different colored flowers in the area. After statistically analyzing the data, it was demonstrated that the bees were attracted to the violet- and yellow-colored flowers and rarely visited the pink and red flowers. The scientist used
   I.   Inductive reasoning   II.   deductive reasoning   III.   the scientific method
a. I only   b. I and II   c. I, II, and III   d. I and III   e. II and III

8. A daughter was born in a family with four sons. What is the probability that a family would have four sons and only one daughter?
a. 1/4   b. 1/8   c. 1/16   d. 1/32   e. 1/64

9. In the rainforest of Ecuador, oil and pollutants are dumped by oil companies into local rivers used by the indigenous people. This practice is an example of
a. toxic colonialism.   b. ecofeminism.   c. environmental parity.
d. anthropocentric worldview.   e. environmental development.

10. Which of the following is an example of a controlled experiment?
a. A botanist exposes three types of plants to varying levels of carbon dioxide to test growth.
b. A zoologist watches the behavioral interactions of a group of monkeys.
c. A researcher examines the toxicity of two pesticides by exposing one fish to both pesticides.
d. A toxicologist treats one group of *Daphnia* with lead and compares the results to another group of *Daphnia* not exposed to lead.
e. A scientist studies the combined impact of cigarette smoking and drinking alcohol by looking at the increase in liver and lung cancers in smokers that drink.

Explanations

1. d. Nitrates are the independent variables. The soil and rye seeds are constants of the experiment. The height and mass are the dependent variables because they are dependent upon the presence of the nitrates.

2. c. The clover must be fixing nitrogen from the atmosphere and enriching the soil thus enhancing the growth of the rye grass. Because the nitrates are lacking in the fertilizer, that pot 1 rye grass has nearly the same increase in height and mass as pot 2 indicates that the clover was fixing nitrogen for the rye seeds to use. The nitrogen-free media would not cause increased growth of the plants just because other seeds were present. Clearly the growth was better in pot 2 than pot 1, therefore the clover seeds could not be inhibiting the growth of the rye grass, nor would crowding enable plants to grow better. There was growth in pot 1, so the nitrogen-free media could not be toxic to the rye seeds.

3. d. Utilitarian conservationists believe in protecting resources for their economic value because utilitarian conservationists wish to use the resources for humans. Its value as wildlife habitat for wildlife, aesthetic value, intrinsic value, and role in the biosphere does not generate economic worth.

4. b. Environmental racism is subjecting people of color to hazards not present for the rest of the population. Therefore, the presence of hazardous waste sites on land owned by Native Americans is toxic colonialism.

5. a. The definition of hypothesis is a testable explanation for an observation.

6. a. The sulfur dioxide levels dropped since 1990. There was not a 50 percent decline in sulfur dioxide emissions from 1980 to 1990. You also cannot tell that the sulfur dioxide levels will continue to drop. The graph does not differentiate between natural and anthropogenic sources. On the contrary, sulfur dioxide emissions are not linked to global warming.

7. a. The scientist was using inductive reasoning by applying the scientific method. He gathered data based upon observation and repeated the experiment in four different fields for replication. He came to the general conclusion that the violet and yellow flowers were favored, thus was using inductive reasoning.

8. d. The probability of events happening simultaneously is the product of their individual probabilities. Therefore, five children, each with a one-half probability of being male or female, results in a one-thirty-second probability of having four sons and only one daughter. $\frac{1}{2} \times \frac{1}{2} \times \frac{1}{2} \times \frac{1}{2} \times \frac{1}{2} = 1/32$

9. a. Toxic colonialism is the release of toxins in undeveloped countries because they do not have the means to fight the injustice.

10. d. Because the toxicologist treats one group of *Daphnia* with lead and compares the results to another group of *Daphnia* not exposed to lead, this is a controlled experiment. The botanist did not have a control, nor did the zoologist. The toxicity of pesticides should have been tested on numerous fish, but separately. The increased lung and liver cancers may not demonstrate the impact of smoking and drinking unless the results are compared to the results from people with those cancers that did not drink or smoke.

Chapter 3 – Matter, Energy, and Life

Key Terms

acid	ecological pyramid	organic compound
ammonification	ecosystem	oxidation
assimilation	element	pH
atom	energy	phosphorus cycle
atomic number	entropy	photosynthesis
base	enzyme	population
biomass	food chain	potential energy
carbohydrate	food web	producer
carbon cycle	heat	protein
carbon sink	herbivore	pyramid of biomass
carnivore	hydrologic cycle	pyramid of energy
cell	ion	pyramid of numbers
cellular respiration	ionic bond	radioisotopes
chemical energy	isotope	reduction
community	kinetic energy	salt
compound	lipid	scavenger
conservation of matter	molecule	species
consumer	nitrification	sulfur cycle
covalent bond	nitrogen cycle	thermodynamics
decomposer	nitrogen fixation	trophic level
denitrification	nucleic acid	
detritovore	omnivore	

Skills

1. Recall introductory chemistry terminology, including elements and atoms.
2. Differentiate between ionic and covalent bonds.
3. Classify acids, bases, and salts based upon their chemical properties.
4. Distinguish between organic and inorganic compounds.
5. Discriminate between the four major groups of organic compounds found in living things.
6. Define cell, and describe cellular structure.
7. Define energy, and identify the different types of energy important to living organisms.
8. Summarize the first and second laws of thermodynamics.
9. Examine the role of the sun in life on earth.
10. Characterize the interaction between photosynthesis and cellular respiration, including their role in the carbon cycle.
11. Identify the ecosystem hierarchy levels.
12. Distinguish between food chains, food webs, and trophic levels.
13. Draw the three types of ecological pyramids.
14. Diagram the nitrogen, carbon, and phosphorus cycles. Include in the diagram the processes that make up the cycles.

Introductory Chemistry

Figure 3.1 Atomic structure

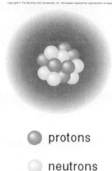

Matter is anything that occupies space and has mass. Matter is made up of elements, substances that cannot be broken down into simpler substances by conventional means. An atom is the smallest particle of an element that still retains the properties of that element. Examples of elements important to environmental science include hydrogen, carbon, chlorine, phosphorus, uranium, and lead. Atoms have positively charged protons and uncharged neutrons in the center, called the nucleus. The mass of protons and neutrons is similar. Rotating around the nucleus in a predictable location are electrons, which are negatively charged. Electrons are much smaller than the nuclear particles. The net charge on an atom is neutral, because the number of protons equals the number of electrons.

protons

neutrons

electrons

The atomic number of an element is the number of protons it contains. The atomic mass of an element is the number of protons plus the number of neutrons. A single element may have several different isotopes, which means although they always have the same number of protons, the number of neutrons may vary. For example Uranium has an atomic number of 92, meaning it has 92 protons. The isotopes of uranium are U-235, U-238, and U-239 and they have 143, 146, and 147 neutrons, respectively. Some isotopes are unstable due to the presence of the neutrons and may emit particles and/or energy. These isotopes are known as radioactive isotopes or radioisotopes.

Take Note: You will need to be familiar with the various elements in living organisms. You should also know which elements have a utilitarian use and may be mined by humans. Knowing which elements are found in the earth's crust is also important. You will need to know which metals are considered heavy metals and understand the damage that they cause to humans and the environment. It is important that you know how specific isotopes are used in science and industry. You should also know the problems associated with their use and the waste generated by their use.

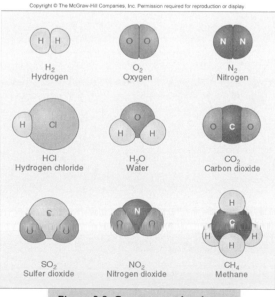

H_2
Hydrogen

O_2
Oxygen

N_2
Nitrogen

HCl
Hydrogen chloride

H_2O
Water

CO_2
Carbon dioxide

SO_2
Sulfer dioxide

NO_2
Nitrogen dioxide

CH_4
Methane

Figure 3.2 Common molecules

A molecule is composed of two or more atoms that exist as a single unit, such as H_2, N_2, or O_2. A compound is a substance composed of different types of chemically bonded elements. For example, water (H_2O), table salt (NaCl), and methane (CH_4) are compounds. All compounds are therefore molecules, but not all molecules are compounds.

Elements form compounds and molecules to have a stable number of electrons. An ionic bond is formed when one element loses or gains electrons to form an ion, or charged particle. The two ions are then held together due to their opposite charges. The atom that gains the electron is said to be reduced and the atom that loses the electron has been oxidized. The reaction is deemed a redox or reduction—oxidation reaction—because it occurs simultaneously. A negatively charged ion is called an anion and a positively charged ion is known as a cation.

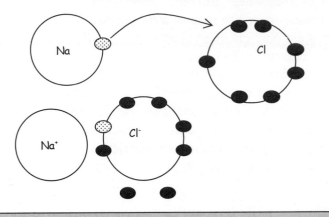

Figure 3.3 Ionic Bond: Na will give up its one electron in its outer shell to Cl to be stable. Na has one less electron than protons, so it now has a plus one charge and is a cation, Na^{+1}.
The Cl atom has gained an electron in its outer shell and has thus become negative by one charge and is the anion Cl^-.
The ions are held together due to their attraction for the opposing charge, forming an ionic bond.

A covalent bond is formed when two atoms share electrons. Some atoms tend to hold electrons more closely to the nucleus and are deemed more electronegative than the atom with which it is sharing the electrons. An example of this uneven distribution of charge can be seen in water, where the oxygen is slightly negatively charged and the two hydrogens are slightly positively charged. In this event, although the molecule is not charged, portions of the molecule are charged because the electrons tend to remain more closely in association with the more electronegative atom (O in this example). These molecules are called polar molecules because they possess opposing charges, and they form hydrogen bonds, because hydrogen is usually the element that is sharing the unequal electron distribution with a more electronegative atom.

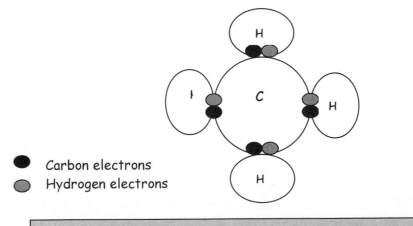

● Carbon electrons
● Hydrogen electrons

Figure 3.4 Methane. The carbon atom shares its four electrons with four hydrogen atoms, thus forming a covalent bond. Neither carbon nor hydrogen is highly electronegative, so the electrons are shared equally between the atoms.

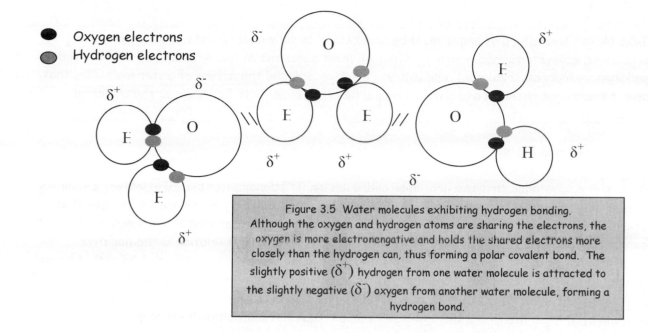

Figure 3.5 Water molecules exhibiting hydrogen bonding. Although the oxygen and hydrogen atoms are sharing the electrons, the oxygen is more electronegative and holds the shared electrons more closely than the hydrogen can, thus forming a polar covalent bond. The slightly positive (δ^+) hydrogen from one water molecule is attracted to the slightly negative (δ^-) oxygen from another water molecule, forming a hydrogen bond.

Typically the formation of chemical bonds requires energy, or is an endergonic reaction. The breaking of bonds is usually exergonic, or an energy releasing reaction. The energy required to initiate a chemical reaction is known as activation energy.

Water

The polar nature of water is extremely important in explaining its properties that are essential for life.

- Water forms hydrogen bonds with other water molecules, other polar molecules and ions. Water has therefore been deemed the "universal solvent," since it can dissolve the polar molecules (such as sucrose) and ions (such as table salt).
- Water bonding to other water molecules is known as cohesion, and creates the property known as surface tension. Surface tension is the resistance of water surface to being penetrated by an object, such as a water strider walking on a pond.
- Water bonding to other polar molecules is known as adhesion, which in conjunction with cohesion results in capillary action. Capillary action is when water is drawn through a channel or vessel. For example, the water flowing through the xylem of a plant is capillary action.
- Water has a high heat of vaporization, which means it takes a lot of heat to convert if from liquid to gas.
- Water has a high specific heat, meaning it must absorb a lot of heat for the temperature to rise.
- Unlike most materials, water expands when it freezes, assisting soil formation due to frost wedging in the crevices of rocks. The water expands, so it is less dense when it is frozen than liquid water. Therefore, ice floats on the surface of lakes, allowing survival of the organisms in the lake even in the dead of winter.
- Water spontaneously ionizes into hydrogen ions and hydroxide ions. These ions remain in equivalent amounts in pure water, and water therefore has a neutral pH.

Take Note: The nature of water must be understood to fully understand water's role in living organisms, ecosystems, and weather. Although most questions on the AP exam will address water pollution, resources, treatment, and use, you must understand the nature of water molecules that give it the unique chemical and physical properties that cause it to be known as the universal solvent.

Acids, Bases, and Salts

An acid is a substance that is a proton or hydrogen ion (H^+) donor when placed in water. Examples of acids include hydrochloric acid (HCl) and acetic acid (CH_3COOH). Bases are substances that release hydroxide ions (OH^-) in water and therefore accept hydrogen ions to form water. An example of a strong base is sodium hydroxide (NaOH). The pH of a solution is the negative logarithm base ten of the concentration of hydrogen ions. The pH scale is from 1–14 with 7 being neutral. Below pH 7 is acidic and above pH 7 is basic or alkaline. Therefore, a substance with a pH of 4 contains 1×10^{-4} (0.0001) H ions and a substance with pH of 6 has 1×10^{-6} (0.000001) ions. The solution of pH 4 is therefore 100 times more acidic than the solution with pH of 6.

When an acid is combined with a base, water and a salt are formed, neutralizing the pH. For example, HCl + NaOH → NaCl + H_2O. A buffer is a substance that allows significant additions of an acid or base without altering the pH of a system because it can accept or donate H ions. An excellent natural buffer is calcium carbonate (limestone or, when powdered, lime).

Take Note: You must be well versed in acids formed from anthropogenic and natural pollutants and their effects on ecosystems to do well on the AP exam. You need to understand the relationship between pH and the H ion concentration. You should be able to understand the principles of pH to be able to discuss acid deposition and how humans can prevent or remediate damage from acids.

Organic Compounds

An organic compound is one that contains carbon. Carbon has the capacity to share four electrons in covalent bonds, and therefore is important in the structure of living organisms. Carbon molecules form large macromolecules because they can be found in long chains and can form ring compounds. Organic molecules found in living organisms include carbohydrates, lipids, proteins, and nucleic acids. See the macromolecule chart in Figure 3.6 for classification of these macromolecules.

Cells

A cell is the smallest unit that can carry out all of life's functions. Cells are separated from their environment by a membrane that regulates entry and exit of materials for the cell. Organelles are small structures within a cell that carry out specific activities for the cell. Cells that have membrane-bound organelles are called eukaryotes and cells that lack membrane-bound organelles are called prokaryotes. All prokaryotes are bacteria. Eukaryotes include both uni- and multicellular organisms such as protists, fungi, plants, and animals.

Figure 3.6 Macromolecule chart

Macromolecule	Elements	Subunits	Role in living organisms	Examples
Carbohydrate	C, H, O	Simple sugars, monosaccharides	Energy storage, cellular structure	Sugar (glucose, sucrose), starch, cellulose
Lipid	C, H, O Nonpolar	Fatty acids, steroids	Energy storage, cell membranes, cell signaling (hormones)	Waxes, fats, oil, phospholipids, some hormones
Protein	C, H, O, N (S)	Amino acids	Enzymes (protein catalysts that speed up chemical reactions by lowering activation energy), structural components, cell signaling (hormones), cell markers	Enzymes, antibodies, hemoglobin, some hormones, actin, and myosin in muscles
Nucleic Acids	C, H, O, N, P	Nucleotides	Nucleic acids—genetic material Nucleotides—cell signaling and energy storage	DNA/RNA—nucleic acids cAMP, ATP—nucleotides

Energy

Energy is the ability to do work. Energy of motion is called kinetic energy, and the energy of an object due to its position is potential energy. Some potential energy is in the form of chemical energy, such as the energy in a candy bar that will be available to your body after you consume it. Energy is measured in work (joules) or heat units (calories). A joule (J) is the work done when 1 kg accelerates 1 m/s/s. A calorie is the amount of energy needed to increase the temperature of 1 g of water 1°C. One calorie = 4.184 J.

Heat is the energy transferred between two objects of differing temperature. Diffuse energy is considered low-quality energy because it is not useful for work. For example, the energy in the ocean is too diffuse to be useful, but the energy in coal is concentrated and is thus considered high-quality energy.

Take Note: The AP exam will contain numerous examples of problems using energy conversions. Students must be adept at these energy conversions. These questions may be found not only in the essay portion of the exam, but conversion problems may also be found in the multiple-choice portion. A basic understanding of energy and thermodynamics is essential to understanding alternative and conventional energy resources.

Thermodynamics

Thermodynamics is the study of energy conversion. There are two laws that govern energy changes. The first law of thermodynamics, also known as the law of conservation of energy, states that energy can neither be created nor destroyed; it can only change from one form to another. This law applies to conventional energy conversions—not nuclear change. In nuclear change matter and energy are converted simultaneously, leading to the coupling of two laws—the law of conservation of matter and energy.

The second law of thermodynamics states that with each energy conversion in a closed system, the energy change proceeds toward entropy, or a state of disorganization. Disorganization is favored in nature because it is more stable than organization. A good example is a stack of soda cans forming a pyramid on a table. The stack is very organized, but unstable as you could easily knock it over. All the soda cans lying on their side on the table are disorganized, but stable.

The Sun

The sun is the principle source of energy for all life on earth. It provides heat that warms the earth and allows life to exist. The sun also provides the photons, packets of light energy, absorbed by pigments in plants, algaes, and cyanobacteria to initiate photosynthesis. Photosynthesis uses water and carbon dioxide in the atmosphere to produce simple sugars, which the plant can then use for energy and can bind in polymers for storage and structural components. The by-product of photosynthesis, oxygen, is released into the environment. About 1 percent of the sun's energy that reaches the earth is converted by photosynthetic organisms into chemical energy. Organisms that create their own food are called producers. Not all producers are photosynthetic. Some are chemosynthetic, using the inorganic materials in the environment to create chemical energy needed to sustain life. An organism that obtains its nutrients from other organisms is called a consumer. The process of cellular respiration occurs in both producers and consumers. Cellular respiration is the breaking of the chemical bonds created during photosynthesis to release the contained energy to use for life's processes. The reactants in photosynthesis are the products of cellular respiration and vice versa; thus, the two processes are inextricably intertwined.

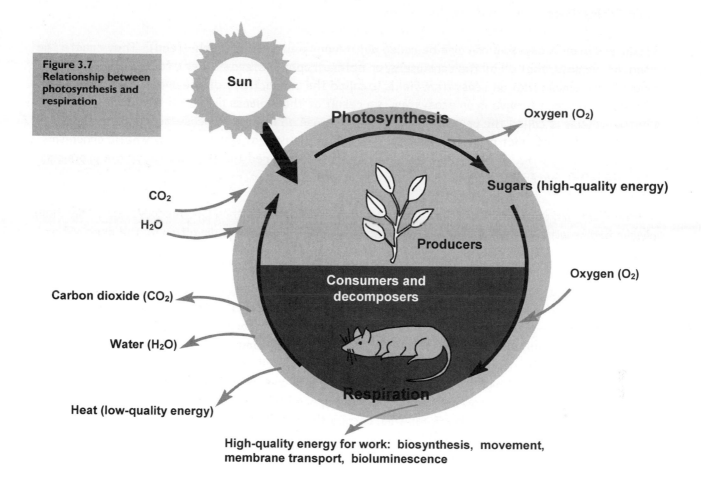

**Figure 3.7
Relationship between
photosynthesis and
respiration**

Ecosystem Hierarchy

The organism is an individual that can carry out all of the characteristics of life. Species is defined as a group of similar looking organisms that can reproduce and produce fertile offspring. For example the gray wolf, *Canis lupus*, is a separate species from the dog, *Canis familiaris*. A population is a group of organisms of the same species living in the same place at the same time—all the gray wolves in Yellowstone National Park. A community is a group of interacting populations, such as the wolves, deer, rabbits, aspens, and birds living in the park. The ecosystem is the complex interactions of the biological community and its environment. The living components are deemed biotic (bio - life) and the nonliving components are abiotic (a—without; bio—life). In the park, the abiotic components that affect the biotic community would be temperature, precipitation, altitude, soil, wind, and so on.

The biosphere is the portion of earth that supports life, including the lower atmosphere and the upper crust.

organism→species→population→community→ecosystem→biosphere

- 22 -

Food Chains, Food Webs, and Trophic Levels

Producers in an ecosystem can also be called autotrophs (auto—self; troph—feed). They create the food, or biomass, that all of the consumers, or heterotrophs (hetero—other), feed upon. The amount of biomass that an ecosystem yields is called the productivity of the area. The number of creatures that can survive in an ecosystem is a result of the biomass produced in the area. Photosynthesis is called the primary productivity because all life in the ecosystem is dependent upon the producers. Net primary productivity (NPP) is the rate at which photosynthetic organisms make sugars minus the sugars used in cellular respiration required for the survival of the producer. The NPP is the material available to the consumers in the ecosystem.

A food chain is a series of links that illustrates how energy moves through an ecosystem. The chain always begins with a producer and the arrows are always placed in the direction of the energy flow. Food chains rarely have more than five links, due to the energy lost at each feeding, or trophic, level. Food chains are overly simplified interactions. In reality, organisms have more than one food source and may be the food source of more than one organism. The resulting linked food chains are called a food web, a much more realistic diagram of feeding and energy transfer in an ecosystem. Herbivores (herb—plant; vore—eat) eat vegetation. Carnivores (carn—meat) eat other consumers. Omnivores (omni—all) ingest both plants and other consumers. Scavengers, such as hyenas or condors, consume the dead carcasses of other animals. A detritovore ingests detritus, the partially broken down leaf litter, plant remains, and dung present in an ecosystem. Detritovores include earthworms, millipedes, and ants. Decomposers are specialized bacteria and fungi that break down organic material in the environment and then absorb the nutrients.

Food Chain   Algae→bluegill→bass→osprey

Figure 3.8 Food web

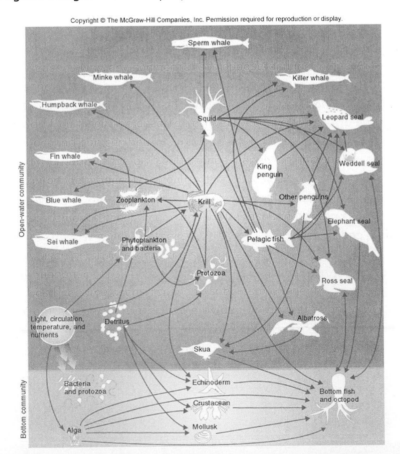

Take Note: You must be able to generate food chains/webs in specific ecosystems. You must also be able to understand energy flow through food webs. The terminology of the different trophic levels will be used in numerous ways on the exam.

Ecological Pyramids

In addition to food chains, there is another way to diagram the passage of biomass and energy in an ecosystem. An ecological pyramid is a schematic representation of the food chain used to illustrate the broad base of producers typically required to support the other organisms in an ecosystem. There are three major types of pyramids.
- Pyramid of numbers
  - Shows the numbers of organisms at each trophic level
  - May be inverted in aquatic ecosystems because of the high reproductive rate of algae and some forests due to few trees providing the biomass for the herbivores

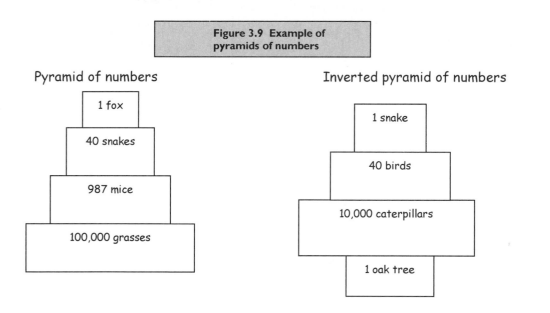

Figure 3.9 Example of pyramids of numbers

Pyramid of numbers

| 1 fox |
| 40 snakes |
| 987 mice |
| 100,000 grasses |

Inverted pyramid of numbers

| 1 snake |
| 40 birds |
| 10,000 caterpillars |
| 1 oak tree |

- Pyramid of biomass
  - o Demonstrates the mass of organisms at each trophic level
  - o May be inverted in aquatic ecosystems due to high reproductive rate of algae capable of feeding a larger biomass of zooplankton

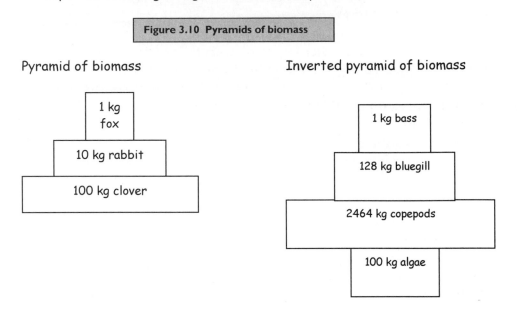

Figure 3.10 Pyramids of biomass

Pyramid of biomass

Inverted pyramid of biomass

- Pyramid of energy
  - o Energy is lost between successive trophic levels due to a variety of reasons
  - o Consumer may not be able to digest all of the organism that it ingested
  - o Consumer may have had to expend energy to catch its food
  - o Consumer may not be able to eat the entire organism that it killed
  - o Consumer loses thermal energy
  - o As a result of these losses, ecosystems lose anywhere from 80–95 percent of their available energy between trophic levels. The average transfer of energy between trophic levels is about 10 percent.

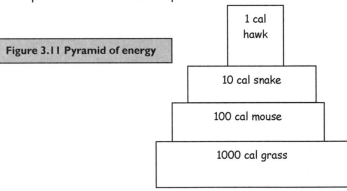

Figure 3.11 Pyramid of energy

Biogeochemical cycles

Because matter is not created nor destroyed, so it must be cycled on earth. The cycling of matter is exceptionally important for moving materials from the biotic to the abiotic portions of the biosphere. These cycles are therefore called biogeochemical cycles. The macronutrient cycles that are most important are the water, carbon, nitrogen, phosphorus, and sulfur cycles. These are cycles, so there is no beginning and no end. Understand the conversion from one form to another and human interferences in these cycles.

Take Note: You must understand the role of the cycled nutrients in living things to be successful on the AP exam. Some questions have required students to understand the implications of humans interfering in matter cycling. Also understand how human interference in the cycles impacts the environment. Be specific and give examples of anthropogenic pollutants and/or environmental damage to answer any essays given. For example, the hydrologic cycle may not have appeared as a diagram, but past essay questions have addressed water diversion projects, impacts of the water cycle on soil, and surface water contamination. All of these questions demand an understanding of the way in which water cycles on earth.

Hydrologic or Water Cycle

Water is released from surface water through evaporation. Evapotranspiration is the loss of water from the leaves of a plant as they exchange gases necessary during photosynthesis. Once in the atmosphere the water molecules undergo condensation, and then precipitation occurs, returning the water to the earth. The water will infiltrate the soil and percolate down into the deeper layers of the soil or it will become runoff. The percolated water may become part of the groundwater, which flows steadily underground toward the ocean. The water may also be used by plants during photosynthesis. The runoff will become part of the surface water, entering lakes or flowing water systems. The hydrolic cycle is powered by the sun and gravity.

Human intervention in the hydrologic cycle includes:
- Ground and surface water depletion
- Ground and surface water pollution
- The clearing of vegetation, particularly in temperate and tropical rainforests, interferes with the cycle by decreasing transpiration

Figure 3.12 Water cycle

Carbon Cycle

CO_2 makes up 0.035 percent of the atmosphere. It is used by plants during photosynthesis and incorporated into carbohydrates. The carbohydrates are then used during cellular respiration, giving off CO_2 into the atmosphere again. Decomposers degrade the remains of plants and animals and return the CO_2 to the atmosphere as well. Much of the carbon on earth is in the form of limestone, a type of sedimentary rock. In the oceans carbon is found dissolved as carbonate and bicarbonate ions. Large amounts of calcium carbonate ($CaCO_3$) deposits are found on the ocean floor, as they are the hard remains of organisms such as mollusks and corals. Heavily vegetated areas are called carbon sinks because they store large amounts of carbon in the biomass. In ancient times these heavily vegetated areas formed fossil fuels, also carbon sinks.

Human intervention in the carbon cycle
- Since the industrial revolution, we have dramatically increased the CO_2 in our atmosphere due to:
  - deforestation, which decreases the plants available for photosynthesis, thus decreasing the uptake of CO_2
  - forests burning and returning the carbon in the biomass of the forest to the atmosphere by releasing it from the sink
  - increased combustion of fossil fuels, releasing carbon from the sinks
- Increased CO_2 in the atmosphere has exacerbated global warming by holding in infrared heat around earth that would normally escape into space.

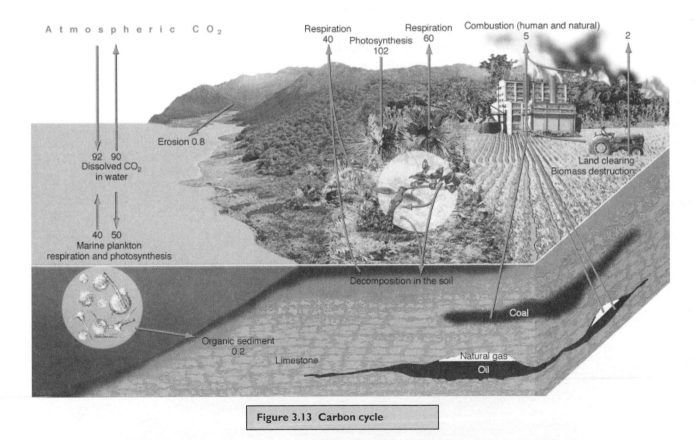

Figure 3.13 Carbon cycle

Nitrogen Cycle

The nitrogen cycle has five major steps. Nitrogen fixation is the conversion of atmospheric N_2 into ammonia (NH_3). Some bacteria, such as *Rhizobium* and some cyanobacteria, are capable of breaking the triple covalent bond found in the diatomic nitrogen gas and combining the nitrogen with hydrogen to form ammonia. *Rhizobium* lives in the root nodules of legumes such as soybeans, alfalfa, peas, and clover. Some plants can use ammonia, but more can use nitrates. The next step is called nitrification. Other bacteria convert the ammonia into nitrites (NO_2^-, not usable by plants) and then a different group of bacteria convert the nitrites into nitrates (NO_3-, which plants can use). Assimilation is when the plant roots absorb ammonia, ammonium ions, or nitrates and make the substances they require for life. Ammonification takes place when the dead nitrogen rich organisms, their parts, or their metabolic wastes are converted to ammonia and ammonium ions by decomposition by bacteria. Denitrification is the process of ammonium ions and ammonia converting back to nitrogen gas, released into the atmosphere. This is a process carried out by yet another group of bacteria. Nitrogen can be an important limiting factor in terrestrial ecosystems because it is easily leached from the soil.

Human intervention in the nitrogen cycle:
- NO is released when fossil fuels are combusted. NO in the atmosphere forms nitric acid, which results in acid deposition (eventually causing terrestrial and aquatic ecosystems to become more acidic) and smog formation.
- N_2O gas is a greenhouse gas and is derived from livestock waste and use of commercial fertilizer
- Nitrogen is removed from an ecosystem if plants are removed
- Nitrogen, when added to aquatic ecosystems from agricultural runoff and municipal waste treatment, induces eutrophication, an enrichment of the ecosystem resulting in an algal bloom that has far reaching consequences
-

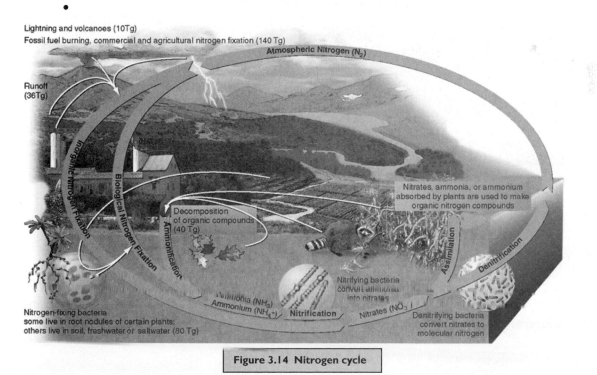

Figure 3.14 Nitrogen cycle

Phosphorus Cycle

The phosphorus cycle is very slow because there is no atmospheric stage. Nearly all of the phosphorus is in living organisms or in rock, because the ions of phosphorus do not dissolve well in water. Most is found in insoluble rock on the ocean floor as a phosphorus sink. The rock containing phosphates must be weathered to release the minerals or removed by mining. The phosphorus is taken up by producers and transferred to consumers. The decay of their bodies releases the phosphorous again. Guano, the phosphorus-rich feces from fish-eating birds, is a part of the cycle.

Human intervention in the phosphorus cycle:
- Phosphate mines that form large pits and result in runoff pollution
- Removing vegetation lowers phosphorus availability in the ecosystem
- Similar to increases in nitrogen, increases in phosphorus leads to eutrophication in aquatic ecosystems

Figure 3.15 Phosphorus cycle

Sulfur cycle

Most sulfur is found in rock as iron disulfide (pyrite) or as mineral salts, like calcium sulfate (gypsum). The sulfur is released primarily by weathering and volcanic activity. The sulfur is taken up by the producers, passed to the consumers, and released again after decomposition.

Human intervention in the sulfur cycle:
- Fossil fuel combustion releases sulfur dioxide, which forms sulfuric acid in the atmosphere and results in acidification of ecosystems, reduced visibility, and human health problems
- Refining of petroleum and smelting releases sulfur compounds
- Coal mining results in sulfur release, which may cause damage to aquatic ecosystems if near surface water
- Large amounts of sulfur dioxide and sulfate aerosols cool the atmosphere because they prevent the penetration of UV radiation

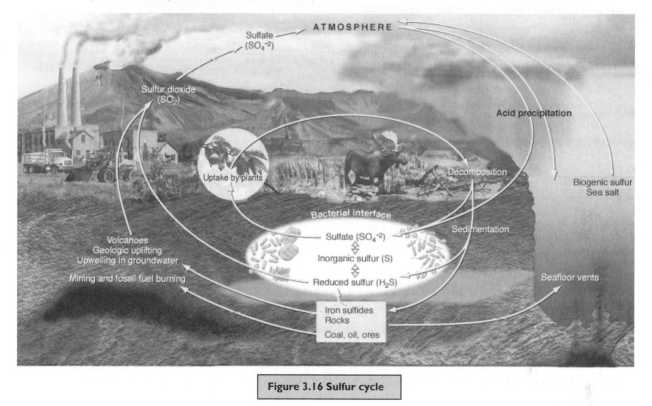

Figure 3.16 Sulfur cycle

Chapter 3 Questions

1. In a pond algae has 10,000 kcal of available energy. If the zooplankton, which consumes the algae, has 1000 kcal of available energy. What is the available energy to the minnow that consumes the zooplankton?
   a. 1,000 J   b. 100 J   c. 200 J   d. 10 J   e. 1 J

2. Nitrogen in the atmosphere is converted into ammonia by the process called
   a. nitrification.   b. denitrification.   c. nitrogen fixation.
   d. ammonification.   e. assimilation.

3. A solution with a pH of 2 is ____ times more acidic than a solution with a pH of 5
   a. 10   b. 30   c. 100   d. 300   e. 1,000

4. Which of the following organisms is an herbivore?
   a. bobcat   b. wolf   c. rabbit   c. eagle   d. snake

5. Plants require phosphorus to form which compound?
   a. simple sugars   b. deoxyribonucleic acid   c. proteins   d. carbohydrates   e. lipids

6. Which of the following is an example of kinetic energy?
   a. water in a reservoir   b. gasoline in a motor   c. wind blowing at the beach
   d. a stretched rubber band   e. a rock at the top of a hill

7. Cellular respiration produces which of the following gases?
   a. CO_2 b. CH_4 c. O_2 d. N_2 e. H_2

Use the paragraph for questions 8 and 9.

In the African savanna, Thompson's gazelles consume short grasses. The gazelles are a food source for cheetahs. The gazelle remains are ingested by hyenas and vultures.

8. What role do the hyenas play in this ecosystem?
   a. herbivore   b. carnivore   c. scavenger   d. decomposer   e. detritovore

9. The organisms described make up a(n)
   a. ecosystem.   b. community.   c. population.   d. species.   e. biosphere.

10. Humans interfere in the hydrologic cycle in all of the following ways except:
    a. building dams that flood ecosystems.   b. channelizing rivers to allow easier boat passage.
    c. transpiration from leaves enters the atmosphere.   d. deforestation in tropical areas.
    e. pumping groundwater for irrigation.

11. An isotope differs from other elements of the same atomic number due to its different number of
    a. protons.   b. atoms.   c. ions.   d. neutrons.   e. electrons.

Use the diagram to answer questions 12-15.

12. Organisms in trophic level II in the diagram would be classified as
    a. autotrophs.   b. decomposers.   c. scavengers.   d. herbivores.   e. carnivores.

13. The organisms at level IV would most likely be classified as
    a. primary consumers.   b. primary producers.   c. secondary consumers.
    d. herbivores.   e. top carnivores.

14. Which of the following statements is true about the pyramid of energy in the diagram?
   a. The energy from level IV will be passed back to level I eventually.
   b. The autotrophs would be found at level IV.
   c. The biomass at level IV is greater than the biomass at level I.
   d. The energy available to each subsequent tropic level continues to increase.
   e. The energy lost to the surrounding environment is thermal energy, which is unrecoverable.

15. The energy available to tropic level IV in the absence of the second law of thermodynamics would be
   a. 0.1 percent.   b. 10 percent.   c. 50 percent.   d. 75 percent.   e. 100 percent.

16. Photosynthesis products used during cellular respiration include
   a. CO_2, water, and sugars   b. water, oxygen, and sugars
   c. carbon dioxide, water, and energy   d. oxygen and sugars
   e. energy and oxygen

17. The only biogeochemical cycle with no atmospheric component is the _____ cycle.
   a. phosphorus   b. nitrogen   c. sulfur
   d. water   e. carbon

18. Which of the following characteristics of water is correct?
   a. Water has low heat of vaporization.
   b. Water contracts as it freezes.
   c. Water is nonpolar and is attracted to nonpolar substances.
   d. Water exhibits cohesion, the tendency to attach to other molecules.
   e. Water spontaneously ionizes into hydrogen and hydroxide ions.

19. Humans interfere in the sulfur cycle by
   a. burning wood for cooking.   b. using inorganic fertilizers.
   c. burning coal to generate electricity. d. using pesticides on cotton farms.
   e. deforestation of tropical forests.

20. All of the following are carbon sinks except
   a. coral reefs.   b. tropical rain forests.   c. coal seams.   d. mangrove swamps.
   e. deserts.

Explanations

1. b. The algae has 10,000 kcal of energy. By assuming that 10 percent of the energy is available to subsequent trophic levels, the zooplankton has 1,000 kcal available and the minnow would have 100 kcal available.

2. c. Nitrogen gas in the atmosphere is converted into ammonia by nitrogen fixation. Nitrification is the conversion of the ammonia in soil into the useful nitrates that can be incorporated into the plants tissues. Denitrification is the formation of nitrogen gas from ammonia. Ammonification is the formation of ammonia from nitrogen containing wastes. Assimilation is the uptake of nitrates by a plant.

3. e. The pH scale is logarithmic. A solution with a pH of 2 has a hydrogen ion concentration of 1×10^{-2}. A solution with a pH of 5 contains 1×10^{-5} hydrogen ions. The difference between the two numbers is 1,000 fold.

4. c. Bobcats, wolves, eagles, and snakes all consume other animals and are therefore carnivores. Rabbits are herbivores because they eat vegetation.

5. b. Simple sugars, carbohydrates, and lipids contain the elements C, H, and O. In addition to those three elements, proteins also contain N. DNA is the only choice that has P as part of its structure.

6. c. Water in a reservoir and a rock at the top of a hill are gravitational potential energy. Gasoline in a motor is stored chemical energy. The stretched rubber band is stored mechanical energy. The wind blowing is energy of motion, or kinetic energy.

7. a. Neither cellular respiration nor photosynthesis produces hydrogen, nitrogen, or methane gas. None of those gases were mentioned as a part of the life-sustaining energy conversions found in photosynthesis and respiration. The choices are between carbon dioxide and oxygen gas. Respiration uses oxygen and produces carbon dioxide as a by-product.

8. c. The gazelle is an herbivore because it eats grasses. The cheetah is a carnivore as it eats the gazelle's meat. No detritovore or decomposer is mentioned in the question. The hyena and the vulture are scavengers, as they clean up dead carcasses.

9. b. Communities are groups of interacting populations. Therefore, a description of the biotic factors is a description of a community.

10. c. Dams interfere with the hydrologic cycle by flooding ecosystems and preventing water flow downstream. Channelizing rivers allow for more rapid water flow, thus interfering with the cycle. Deforestation removes trees and thus interferes with transpiration and absorption. Pumping groundwater interferes with natural aquifers. Only transpiration is a natural occurrence and is therefore the answer.

11. d. An isotope of an element contains a different number of neutrons. The number of protons defines the element. An atom is the smallest particle that exhibits the characteristic of an element. An ion is a charged atom due to gain or loss of electrons. The number of electrons equals the number of protons.

12. d. The organisms in level I are producers, so level II must therefore be herbivores.

13. e. Level I are the producers in the ecosystem. Level II is comprised of herbivores, thus level III must be carnivores. Level IV feeds upon level III, so they must be top carnivores.

14. e. According to the second law of thermodynamics, all energy proceeds toward entropy, or disorganized thermal energy. This pyramid is not inverted therefore c could not be correct. Energy cannot be created nor destroyed it can only change form therefore b is untrue.

Producers only make up the first trophic level so *d* is untrue. Energy in systems cannot cycle, thus *a* is also untrue.

15. e. The energy available to the tropic level IV in the absence of the second law of thermodynamics would be 100 percent because no energy would be lost as thermal energy.

16. d. Photosynthesis uses carbon dioxide and water so they are not answers. Sugars are required for respiration, so *e* is not correct.

17. a. All of the other biogeochemical cycles have some part of the cycle where the elements exist as a gas in the atmosphere.

18. e. Water has a high heat of vaporization, expands as it freezes, and is polar. Cohesion is the tendency for water to stick to itself.

19. c. Coal combustion releases sulfur oxides. Wood combustion and deforestation interfere with the carbon cycle, and inorganic fertilizers interfere with nitrogen and phosphorus cycles. Pesticides do not generally impact the biogeochemical cycles.

20. e. Coral reefs, tropical rain forests, coal seams, and mangrove swamps are carbon sinks because they store a tremendous amount of carbon. Deserts have little biomass and thus store little carbon.

Chapter 4 – Evolution, Biological Communities, and Species Interactions

Key Terms

abundance	evolution	predation
climax community	exotic species	primary productivity
coevolution	fire climaxed communities	primary succession
commensalism	fundamental niche	principle of tolerance
competition	generalist	realized niche
complexity	habitat	resilience
convergent evolution	interspecific competition	resource partitioning
divergent evolution	intraspecific competition	secondary succession
diversity	keystone species	specialist
ecological niche	natural selection	stability
ecological succession	mutualism	symbiosis
ecotone	parasitism	territorality
edge effect	pathogen	tolerance limits
environmental indicators	pioneer species	

Skills

1. Relate tolerance limits to the survival of an organism.
2. Examine the concept of natural selection and understand how adaptation plays a role in evolution.
3. Distinguish between niche and habitat. Compare and contrast the ecological, fundamental, and realized niches an organism may occupy. Explain the following species interactions and give examples of each: predation, parasitism, mutualism, commensalism, and coevolution.
4. Define keystone species and give examples.
5. Differentiate between Batesian and Müllerian mimicry.
6. Relate competition to the survival of individual organisms and to species survival. Characterize resource partitioning as a way to avoid competition.
7. Relate the primary productivity of an ecosystem to its diversity.
8. Characterize an ecotone. Relate their significance in species diversity.
9. Summarize ecological succession. Differentiate between primary and secondary succession.
10. Assess the potential damage to an ecosystem and indigenous species when an invasive exotic species is introduced.

Take Note: It is of paramount importance that you understand the basic ecological precepts presented in this chapter. You can expect to have to explain the role of certain species in the environment. Species that are keystone or exotic dramatically influence their environments. You must be able to explain species interactions, discuss human impacts on species, and give specific examples of these species. Understand ecological succession and how climax communities are determined by the climate and soil in different biomes.

Law of Tolerance

The law of tolerance states that the factor in lowest supply in an environment will be the factor that determines the survival of an organism. Tolerance limits are the maximum and minimum values that encompass an organism's ability to survive. Occasionally a single factor determines the tolerance limit for a species; more often, several factors function together to determine survival. A tolerance curve illustrates the range for survival of a species. The optimal range is the set of conditions that result in abundant numbers of the species. In the zone of physiological stress, the old, sick, and young are culled from the population. The zone of intolerance is the range of conditions that will result in death of even the most hearty of the species.

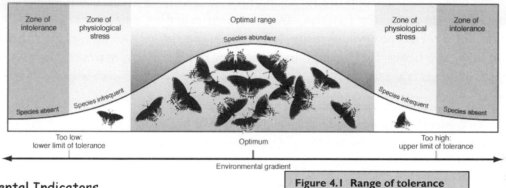

Figure 4.1 Range of tolerance

Environmental Indicators

Environmental indicators are species whose presence indicates a certain environmental parameter is present in a community. Environmental indicators are a direct, visible key to determine the condition of the environment. The presence of cattails in Florida's Everglades is a sign of high levels of phosphates in the water. Cattails are therefore indicative of phosphate pollutants and serve as an environmental indicator of the pollution. In studies of stream benthic macroinvertebrates, mayfly, dobsonfly, and caddisfly larvae are indicative of excellent water quality, whereas the presence of blood worms, leeches, and pouch snails are indicative of very poor water quality.

> **Take Note:** You must be able to define environmental indicator and explain their significance to ecosystems. You should also be able to give specific examples of indicator species, and explain what their presence or absence implies.

Natural Selection and Evolution

Charles Darwin did not devise the theory of evolution; he proposed a mechanism by which natural selection occurs. Organisms more suited to their environment survive to reproduce. Those organisms not as suited do not survive, therefore, they do not reproduce. To a biologist, the term fitness refers to an organism's ability to reproduce successfully. Therefore, survival of the fittest means survival and reproduction of those most suited to the environment. In this fashion, favorable adaptations are passed to subsequent generations and unfavorable adaptations are phased out of the population of organisms. An individual organism cannot evolve; only populations can evolve. Natural selection can be attributed to the alteration of the gene pool in these populations.

The ultimate source of variation in genes is mutation. If a mutation is favorable, the organism becomes more suited to the environment. If the mutation is unfavorable, the trait will be unlikely to enhance the survivability of the species. The environmental conditions that result in natural selection cause selective pressure to be placed upon organisms. The pressure selects for those most suited to survive in the environment. Some selective pressures include physiological stress, predation, competition, and chance. For example, the Galapagos finches underwent adaptive radiation. When an ancestral finch landed on the islands from the mainland of South America, there were numerous niches to fill. Over time, the finches developed adaptations to allow them to fill the available niches.   The adaptations allowed the finches to radiate.

Divergent evolution is when organisms become less alike over time but share a common ancestry. An example is the arm of a human and the forelimb of a dog. The bone structure is similar but each limb is adapted for its particular function. In convergent evolution, selective pressures cause two organisms lacking a recent common ancestor to become more similar. An example is the wing of a bird and the wing of a butterfly. Because both fly, they require wings; however, the structure of the two types of wings is dissimilar.

Take Note: Organisms cannot evolve because they want to evolve in a certain pattern, nor can organisms pass on acquired characteristics. Organisms may possess a mutation that makes them more suited, and because they possess that mutation they can survive to reproduce, thus making their offspring more "fit." Ensure that you are familiar with the concept of natural selection and be able to give examples of selective pressures.

Niches

A habitat is the place where an organism lives out its life. The ecological niche is the role it fulfills in an ecosystem. The niche is controlled by the abiotic and biotic conditions of the ecosystem. An organism can occupy its fundamental niche if it is not experiencing competition, a situation highly unlikely in nature. The fundamental niche is the full range of resources or habitat it could use in the absence of competition. The realized niche is the range of resources and habitat an organism occupies due to the constraints of competition. For example, the arrival of the exotic lizard species, the Cuban brown anole, has displaced the native green Carolina anole in parts of Florida. The Cuban anole resides close to the ground, and the green anole lives in the upper layers of vegetation instead of the entire vegetated area that they would occupy in the absence of the Cuban anoles.

Species that can eat a wide variety of foods and live in a great variety of habitats are called generalists. Examples of generalists include rabbits and raccoons. A species that occupies a very narrow niche is called a specialist. Koalas are specialists because they eat only eucalyptus.

The Law of Competitive Exclusion states that two organisms cannot occupy the same niche for an indefinite period without one eventually migrating, dying, or undergoing resource partitioning. The classic example is growing two species of paramecium together in culture. One species was able to use the resources more efficiently and rapidly multiply and the other species died out, as shown in figure 4.2. Occasionally some species appear to occupy the same niche with both surviving, but resource partitioning is usually occurring. Resource partitioning is the use of the resource in a slightly different way or at a different time to allow both species to use the same resource. For example, warblers use different layers of the forest, and yet all use the same forest resources.

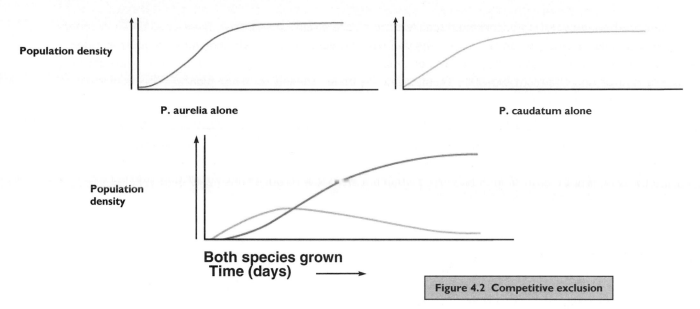

Figure 4.2 Competitive exclusion

Species Interactions

Predators are organisms that feed on other living organisms. The predator feeds upon a prey species. For example an osprey is a predator that preys upon small fish. Predators rely on a strong sense of smell, speed, or stealth to catch their prey. Prey have evolved a variety of mechanisms to avoid predation, including camouflage, noxious taste, spines or thorns, and speed. Some species have evolved camouflage by mimicking vegetation, larger animals, or even eyes. Batesian mimicry is observed when a species evolves to mimic a harmful or noxious species for protection. The Viceroy butterfly is tasty to birds, whereas the Monarch butterfly is noxious. That the mimic resembles the toxic model is protective for the Viceroy butterfly. Müllerian mimicry is based upon the fact that two species may resemble one another when both are toxic. Both species are protected because predators will avoid interacting with them. An example of Müllerian mimicry is that most wasps have yellow and black stripes to demonstrate to predators that they are dangerous.

A special type of predation is parasitism. A parasite is an organism that benefits from a relationship, and the host is the organism harmed by the relationship. Parasites can live on or in the host species. Examples of parasites include ticks, mosquitoes, tapeworms, and flukes. Pathogens, disease causing microbes such as bacteria or fungi, may also be considered parasitic.

Symbiosis is the relationship between two organisms. Parasitism is an example of a symbiotic relationship. Two other types of symbiosis are mutualism and commensalism. Mutualism is an interaction where both species benefit. An example would be lichens or mycorrhizae. Lichens are an interaction between a fungus and an algae or cyanobacteria. The fungus provides shelter and moisture to the algae, and the algae provides sugars to the fungus. Mycorrhizae are interactions between a plant's roots and a fungus. The plant gains surface area for absorption, and the fungus receives sugars in return. Commensalism is when one species is benefited and the other is neither harmed nor benefited. An example is barnacles on a whale. The whale is not damaged or helped by the presence of the barnacle, but the barnacle is given more feeding opportunities by traveling on the body of the whale instead of remaining attached to a stationary surface.

Coevolution is evolution occurring because species are exerting selective pressures on each other. An example of coevolution is that moths pollinate flowers that are strong smelling, light in color, and open at night because most moths are nocturnal species. The animal gains a food resource in the nectar and the plant benefits from having its pollen spread to other plants. This is also an example of mutualism.

> **Take Note:** You must be able to explain advantages and disadvantages of the species interactions. Also, be able to recognize the type of interaction when presented with a scenario, and be able to give examples when presented with questions regarding species interactions.

Keystone Species

A keystone species is a species whose relative abundance may not reflect its importance to an ecosystem. Many keystones are top predators. Gopher tortoises dig burrows in soft sand. Numerous other species use the burrows, including rattlesnakes, foxes, burrowing owls, armadillos, frogs, and insects. Gopher tortoises are therefore extremely important in the biological community, and their removal can interfere with the survival of the other species that depend upon their burrows for protection.

> **Take Note:** You must understand the importance of keystone species to the environment that they inhabit, and be able to give examples and explain their role in community interactions.

Competition

Organisms compete for food, water, shelter, space, mates, and any other component they require for life. Competition between organisms of the same species is known as intraspecfic competition, and competition between different species is known as interspecific competition. Territoriality is a form of intraspecific competition where organisms maintain their own area for mating, food sources, and even nesting sites. Territoriality has the additional benefit of decreasing environmental degradation because species are usually spread out.

Exotic Species

Exotic species are organisms that are not native to an area but are introduced. They may out compete endemic (native) species, lack predators, or damage ecosystems. If they are not held in check in the new environment, they are known as invasive exotic species. These species can cause tremendous amounts of environmental damage, as well as be expensive to eradicate once they establish successful populations. Characteristics of areas that make them more susceptible to invasion by exotics include tropical/semi-tropical climate and high levels of tourism. Examples of exotic species include water hyacinth, zebra mussels, and fire ants.

Community Properties

Primary productivity is the rate at which biomass is accumulated in an ecosystem. The energy left over after cellular respiration is the net primary productivity (NPP). Areas that have high NPP include marshes, coral reefs, and tropical rain forests. Areas with low NPP include deserts, the tundra, and open ocean.

The more complex an ecosystem is, the more likely it is able to withstand damage. Ecosystem diversity is the number of different species occupying different trophic levels. An ecosystem that has great diversity is more stable and therefore more resilient in the event of a disruption.

Boundaries between adjacent ecosystems have tremendous diversity because the two ecosytems transition into each other in an area known as an ecotone. Therefore, species from each ecosystem may survive in the ecotone, resulting in tremendous diversity known as the edge effect. A good example of an ecotone is an estuary. Organisms from both fresh- and saltwater systems can survive in the estuary, thus making it a diverse ecosystem with high NPP.

Succession

Succession is the orderly change of species in an ecosystem over time. Primary succession occurs when an area is newly exposed and has previously not had a biological community. The area lacks soil, and soil must be formed prior to invasion by plant species. Examples of primary succession would be a boulder falling on a mountainside or volcanic flow hardening into bare rock. Pioneer species are the first to grow in an area. Lichens can grow on bare rock and even facilitate the formation of soil by releasing carbonic acid, which breaks down rock into soil. They are the pioneer species in primary succession.

The process by which an early species makes subsequent species more likely to be successful is facilitation. Lichens that contribute to the formation of soil facilitate future plant growth. The lichens and mosses that are pioneer species give way to grasses and shrubs. The next community is the appearance of conifers. Finally, a community of broadleaf hardwoods appears in the site. Secondary succession takes place after an original biological community is damaged and the area undergoes regeneration.

Secondary is much more rapid than primary succession due to the presence of soil. An example of secondary succession would be recovery of an area after a fire or forest regeneration in an abandoned field. Grasses and weeds are the pioneers in secondary succession. Communities continue to change during succession until the final community the ecosystem can support is established. The final community that resists change is known as a climax community. If fire maintains the ecosystem then they are called fire climax communities. These communities regenerate quickly and are able to maintain their stability due to moderate, fairly frequent disturbances.

Take Note: It is important that you understand the process of succession. The final climax community of an ecosystem will be evident in each biome. Also, know the role of fire in maintaining ecosystems.

Chapter 4 Questions

Use the following choices for questions 1-4.
    a. commensalism
    b. mutualism
    c. competition
    d. predation
    e. parasitism

1. aphids make a secretion called honeydew that ants feed upon, and the ants protect the aphids from predators
2. an epiphytic orchid living on an oak tree
3. dolphins chasing mullet in a bay
4. schistosoma worms in a human's blood vessels

5. Primary productivity is highest in which of the following ecosystems?
   a. desert   b. grassland   c. open ocean   d. boreal forest   e. estuary

6. A glacier has retreated from a valley in Glacier National Park, leaving a pile of rocks known as moraine. The type of succession expected in this ecosystem would be
   a. primary.   b. secondary.   c. transitional.
   d. equilibrium.   e. introduced.

7. Which of the following is an example of Müllerian mimicry?
   a. distasteful Monarch and harmless Viceroy butterflies look alike
   b. a moth that resembles a wasp
   c. all species of poison arrow frogs have vivid coloration with black markings
   d. a fly that is fuzzy and brown like a honey bee
   e. a kingsnake has coloration similar to the dangerous coral snake

8. Yellow-headed and red-winged blackbirds live in marshes. Both species can nest in the emergent vegetation in the marsh. If both speices are present, the red-winged blackbirds are pushed into the shallower, drier areas on the outer limits of the marsh. What niche are the red-winged blackbirds using when the yellow-headed blackbirds are in the marsh?
   a. classical   b. fundamental   c. competitive
   d. realized   e. ecological

9. All of the following are adaptive responses to decrease intraspecific competition except
   a. tadpoles are aquatic herbivores and frogs are semiterrestrial carnivores.
   b. maple seeds use their flat wings to move away from the parent plant.
   c. butterflies consume nectar whereas caterpillars ingest leaves.
   d. floating crab larvae and benthic adult crab.
   e. tigers marking their territory.

10. Nocturnal owls minimize competition with diurnal hawks even though they share the same food resource by using
   a. parasitism.   b. resource partitioning. c. coevolution.
   d. intraspecific competition.   e. commensalism.

11. The peppered moths of England underwent directional selection from a light version to a much darker version after the industrial revolution caused tree bark to darken because of the air pollution. Which of the following statements explains the shift?
   a. The presence of the pollutants induced a mutation in the moths, shifting them to a darker coloration.
   b. The organisms were able to mutate to adapt to the new environment.
   c. The new environment caused the darker moths to predominate because the lighter ones were consumed by birds before they could reproduce.
   d. Natural selection caused the dark moths to outcompete the light moths for resources.
   e. The competition between the moths for mates caused the lighter moths to become more rare.

12. Which of the following is true of vegetation during succession?
   a. Grasses predominate in the midsuccessional stages.
   b. Plant biodiversity is greatest in the early successional stages.
   c. The pine forest community facilitates the growth of the hardwoods because they acidify the soil.
   d. Hardwoods will tolerate a few scattered cedars in a late successional forest.
   e. Lichens hinder the growth of grasses as they prevent soil formation when they thickly cover rocks.

13. The flying phalanger, an Australian marsupial, and the North American flying squirrel, a placental mammal, both possess flaps between their front and rear legs to glide for short distances between trees. These animals illustrate the concept of
   a. coevolution.   b. homologous structures. c. convergent evolution.
   d. adaptive radiation.   e. analogous structures.

Questions 14–17 refer to the food web of the chapparal shown below.

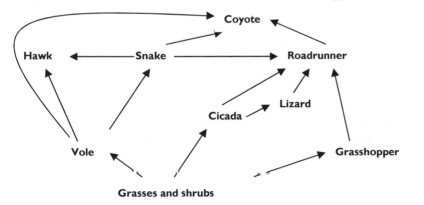

14. Which animal occupies the highest trophic level?
   a. hawk   b. snake   c. roadrunner   d. coyote e. lizard

15. The available energy will be the highest in
a. grasses and shrubs   b. voles   c. grasshopper   d. roadrunner   e. hawk

16. If the population of lizards were to increase, which of the following would occur?
a. The snake and hawk populations will decline.
b. All voles may experience a population increase.
c. The cicada population would increase.
d. There will be a increase in the grasshopper population.
e. The roadrunner population will decrease.

17. The organisms shown form a(n)
a. ecosystem.   b. biome.   c. tolerance range.   d. community.   e. population.

18. The African elephant is known to push over trees in the African savanna. The trees would outcompete the native grasses that many species rely upon without the elephant's intervention. The African elephant's role in the ecosystem is as a(n)
a. keystone species. b. exotic species. c. predator.   d. top competitor.e. parasite.

19. Which of the following would you expect to be the pioneer species on an abandoned farm?
a. hardwood trees   b. pine trees c. lichens d. grasses and weeds   e. shrubs

20. Exotic species damage in ecosystems because they
a. are specialists that die out fairly easily.
b. compete with native species for the same resources.
c. have numerous predators that cause an increase in the number of predators.
d. carry diseases that infect native species.
e. engage in mutualistic relationships with native species.

Explanations
1. b. Both species benefit, so the relationship is mutualistic.
2. a. Epiphytes live on trees, making it a commensal relationship; they do not parasitize them.
3. d. Dolphins are predatory mammals that are carnivores.
4. e. Schistosoma worms are parasites in humans.
5. e. Estuaries have high primary productivity because they are ecotones. The presence of shallow, brackish allows a great variety of flora, and thus more animals. Deserts, grasslands, and boreal forests lack the precipitation necessary for high productivity. Open ocean is too vast and deep to be very productive.
6. a. Transitional, equilibrium, and introduced succession are not terms used in ecology. The example is primary succession, because there is only rock with no soil. Soil must be formed to establish a community.
7. c. All examples are Batesian mimicry except the frogs. Batesian mimicry is when a non toxic mimic is patterned after a toxic model. Müllerian mimicry is when both species are unpalatable and have evolved to be similar. Therefore, that all poison arrow frogs are vividly colored with black markings makes them have Müllerian mimicry.

8. d. The red-winged blackbirds are filling their realized niche, as they are driven from their fundamental niche by the yellow-headed blackbird. The terms classical and competitive niche are not used. The ecological niche is the role played in the community.
9. e. Territorality is an intense form of intraspecific competition. The other choices are mechanisms by which offspring can avoid direct competition with the parent organism.
10. b. Although hawks and owls both feed primarily on rodents, they do so at different times of the day. Therefore, they use resource partitioning allowing for use of the same resource.
11. c. Survival of the fittest explains that the most suited (the darker moths) would predominate as the other moths die out because they are eaten before they can reproduce. Mutations are random and are not induced by the presence of a chemical, nor because the organisms decide to mutate. There is no evidence in the question that the dark moths are able to outcompete the lighter moth for resources or mates.
12. d. The hardwoods will tolerate some conifers in a late successional forest. Grasses would predominate in the early successional stages, and biodiversity is greatest later in succession. A pine forest community inhibits the growth of the hardwoods because they acidify the soil, and lichens facilitate grass growth because they help to make soil from bare rock.
13. c. The flying phalanger and the flying squirrel evolved a similar characteristic due to similar selective pressures.
14. d. The coyote is a quarternary consumer when it eats the roadrunner, a tertiary consumer. The hawk, snake, and lizard are secondary consumers.
15. a. Grasses and shrubs. In all ecosystems the autotrophs would have the greatest energy.
16. d. The grasshopper population will increase. The snakes and hawks would likely increase slightly, since the voles might increase due to a lack of competition with the cicada.
17. d. Community. The organisms form a community because it is a group of interacting populations of animals.
18. a. The African elephant benefits the other species as well as itself by preserving the grassland, so it is a keystone species.
19. d. Secondary succession would take place on an abandoned farm, so soil would be present. Therefore, the pioneer species would be the grasses and weeds.
20. b. Exotic species frequently are generalists that easily compete with native species for resources. They rarely have predators in their new environment. Causing disease is not their primary impact. If they engaged in mutualistic relationships with native species they would not be harmful to the native species.

Chapter 5 – Biomes: Global Patterns of Life

Key Terms

barrier islands	epilimnion	taiga
benthos	estuary	temperate forest
biome	grassland	thermocline
bog	hypolimnion	tropical rainforest
chapparal	mangrove	tropical seasonal forest
cloud forest	marsh	tundra
coniferous forest	pelagic	vertical zonation
coral reef	phytoplankton	wetland
deciduous forest	savanna	
desert	swamp	

Skills

1. Outline the temperature and precipitation conditions in each biome.
2. Evaluate the adaptations of the flora and fauna to each biome.
3. Assess the damage that humans do to the various biomes. Propose mechanisms to repair or avoid the damage in these biomes.
4. Examine the characteristics of each of the biomes.
5. Summarize the variety of marine ecosystems based upon depth.

Take Note: Past AP exams have asked questions about specific biomes. For example, they have asked about deciduous forests and the tundra in essays. You should focus on adaptations of flora and fauna for specific biomes, biodiversity in the biomes, and human disruptions to the biome. Know the impact of the climate and soil on the biome's biodiversity. Be able to explain why a biome may be easily damaged and/or slow to recover from damage. Any impacts of humans on the biome must be explained, and you may be asked to propose policy, suggest preventive measures, or suggest ways to mitigate the already present damage in any of the biomes studied.

Terrestrial Biomes

Biomes are regions characterized by a particular climate, soil, and vegetation. The two most important aspects of the climate that control the type of biome are precipitation and temperature. The latitude affects the location of biomes, because as one moves away from the equator the temperature cools. The tropics are between 23° N (Tropic of Cancer) and 23° S (Tropic of Capricorn). The temperate regions fall between the tropical regions and the polar regions (66.6 ° N and S). Altitude also affects biome location, as the vegetation changes as the altitude increases, which is known as vertical zonation.

Major World Biomas.

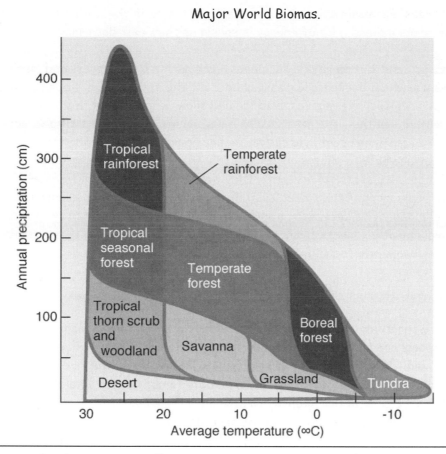

Take Note: The diagram above illustrates the important link between temperature and precipitation in determining the type of biome present in an area. You must understand that both features of the climate determine the biome.

Plant Adaptations

Plants are stationary, therefore they cannot move or migrate during adverse weather conditions. The plants have evolved adaptations that will allow them to survive in various locations. Plants are most frequently affected by precipitation, so many plant adaptations have evolved to deal with a lack of moisture in the environment.   For example, plants in tundra must be low to the ground to retain as much heat as possible. They must also be adapted to low-moisture conditions and a short growing season. Similarly, desert plants must be able to lose heat and yet still retain water, because they, too, live in arid regions. For example, succulents like cactus conserve water in their stem.

The stem also must photosynthesize, because they have no leaves to decrease water loss via transpiration. The modified leaves are called spines, and they protect the plant not only from predation, but from sunlight. They also only open their stomata at night for CO_2 uptake so they do not transpire as much as they would during the day. The broadleaf evergreens keep leaves all year to increase photosynthesis and to allow them to lose heat in areas such as the tropical rain forest or in the southern United States. An example of a broadleaf evergreen is the live oak. The broadleaf deciduous trees, such as maples or oaks, lose their leaves in the winter to decrease water and heat loss. These trees expend a lot of energy in spring as they exit dormancy.

Most conifers are evergreens, but some are deciduous, such as the bald cypress and larch. The conifers possess needles, which decrease water loss because the stomata are hidden within the needle shape. In colder regions, the needles shed ice and snow, which allows the tree to avoid breaking during the winter months. The needles also have a thick, waxy coat that decreases water loss. Evergreens are frequently present in areas where precipitation is a limiting factor. They have needles all year, so they are able to photosynthesize early on in the growing season, which is essential when a short growing season is present. They would be unable to make enough food to survive if they had to replace all of their needles at once because the growing season is so short. Some plants exhibit allelopathy to decrease competition. These plants secrete toxins into the surrounding soil to prevent other plants from growing nearby, thus conserving their resources. Allelopathic plants are frequently found in dry climates, such as the chaparral or deserts.

Fire

Many terrestrial ecosystems rely on the intermediate disturbance of fire to maintain ecosystem homeostasis. The systems are adapted to fire and require periodic burning for regeneration. When fires are extinguished and the ecosystem builds up lots of woody debris, the fire may extend from a relatively harmless ground fire that only consumes the duff (limbs, leaves, and other debris) on the forest floor into a huge canopy fire. Canopy fires will consume even the most hearty of trees as the entire tree will be damaged, usually killing the tree, thus destroying the forest. When only a surface fire is present, the animals can usually escape the blaze. The trees in the fire-based ecosystem frequently have very thick bark that readily resists burning or burns to protect the inner layers of vascular tissue. The fires that consume the duff frequently leave space on the forest floor for new plants to grow. The fires also return nutrients to the soil. Therefore, the fires rejuvenate the area. Many species of conifer even require fire to open their cones, because the newly burned forest will decrease competition for nutrients. Such cones are deemed serotinous and include jack pine, pitch pine, lodgepole pine, and sand pine. Deciduous trees are more susceptible to damage during their growing season.

See the chart at the end of the chapter for a summary of the terrestrial biomes.

Aquatic Biomes

Aquatic biomes are basically split into two groups, freshwater and marine. Freshwater biomes include rivers, streams, lakes, ponds; and all inland wetlands including swamps, bogs, marshes, and fens. Marine biomes include coral reefs, coastal marshes and swamps, estuaries, barrier islands and the open ocean. Many organisms are found in all of the biomes. They can be generalized into groups. The plankton are organisms that float or swim weakly. The two types of plankton are phytoplankton (phyto—plant; plan—wandering), such as cyanobacteria and algae, and zooplankton (zo—animal; plan—wandering), herbivores and carnivores. The phytoplankton are the majority of the producers on earth. The photosynthetic organisms live in the portion of the water that sunlight can penetrate, or euphotic (eu—true; photo—light) zone. The zooplankton can be single celled or larger organisms like a jellyfish. Nekton can swim and are consumers. Sharks, tuna, bass, and perch are all nekton. The benthic organisms live on bottom of the aquatic biome. They are often decomposers and detritovores. Many of them are burrowers including worms and clams. Some are sessile (non-moving) like barnacles and oysters. Some move around on legs like lobsters and crabs. Many of these organisms filter feed, thus functioning to remove pollutants from water.

The most important limiting factors in aquatic biomes are nutrients, dissolved oxygen (DO), temperature depth, and turbidity. The nutrients are required by the autotrophs, and they in turn, provide the DO. The turbidity (cloudiness) determines how deeply the sunlight can penetrate to permit the growth of the autotrophs. Many species have a specific range of tolerance to temperature, so those species will only be found in certain areas of earth. DO also is affected by temperature and is higher in colder water. The depth is important because it impacts light availability and temperature.

Marine Biomes

The majority of life in the ocean is located around the continents, because the nutrients, such as nitrogen and phosphorus, from the land wash into the ocean as runoff.   Ocean currents can carry the nutrients offshore.   The euphotic zones have the majority of the organisms in the ocean because that is the location of the autotrophs. The euphotic zone also has the highest level of DO. Deep ocean areas depend on the death and decay of upper-level organisms falling to the depths as marine snow. Other deep-ocean organisms live around volcanic vents where chemosynthetic organisms form the basis of the food chains. The temperature of the ocean decreases with depth. Shorelines are littoral zones and are highly productive. The area between high and low tides is called the intertidal zone. The open ocean, or pelagic zone, is typically low in nutrients and does not possess the biodiversity present in the other marine biomes.

Coral Reefs

The coral reef is the ocean's equivalent to the tropical rain forest. Coral reefs require relatively shallow, clear, warm waters because coral is a mutualistic relationship between a carnivorous cnidarian and an autotrophic alga. The coral polyps protect the algae within its tissues, and the algae photosynthesize to provide food to the coral. Coral secretes a shell of calcium carbonate that forms limestone rocks. The reef is home to countless species of sponges, cnidarians, worms, echinoderms, crustaceans, and fish. Reefs are subject to damage by any human activity that would restrict their sunlight. For example, increased nitrogen or phosphorus could cause an algal bloom, which could block sunlight to the reef. Any runoff, such as that containing sediment, that would increase the turbidity of the water would adversely affect the reef. Any tropical fish harvest or even coral harvest will affect the reef's productivity and ultimately biodiversity.

Mangrove Swamps

Mangroves are tropical trees that grow along coastlines. They function to catch sediment, remove nutrients, and serve to protect coastlines by preventing erosion. They also have a tremendous amount of biodiversity associated with them, as they serve as aquatic nurseries and nesting areas. Their leaves fall into the water and sink to the bottom, where they serve as food for a variety of detritovores. The food chain is ideal for this area to be used as a nursery for a variety of pelagic fish and numerous invertebrates. Mangroves serve as critical nesting areas for numerous species of birds, including many wading birds like herons and egrets, as well as pelicans. Humans threaten these areas by cutting down the mangroves for lumber and charcoal. They also will convert the swamps to aquaculture areas to raise saltwater species. The areas are also decimated for urbanization and coastline development.

Estuaries

Estuaries are areas where fresh water empties into salt water, resulting in an area with varying levels of salinity, called brackish water. Estuaries have tremendous biodiversity as they are an ecotone with both fresh- and saltwater species present. Estuaries help stabilize shorelines and protect inshore areas. These areas are nutrient rich because the water that flows into them from the land brings in large amounts of nutrients.   Many grasses and emergent vegetation live in estuaries. The water may also be turbid as a result of high amounts of sediment flowing from land, both from natural and anthropogenic (agriculture, construction) sources. The flowing water may also contain numerous pollutants, such as oil, nitrogen, and phosphorus, and fecal contaminants. The estuary has high NPP, and thus serves as a nursery for invertebrates such as shrimp, oysters, and clams as well as for numerous nekton species. Humans use estuaries for recreation (fishing, hunting, boating) and for commercial fishing and aquaculture.

> **Take Note:** A previous essay question on an AP exam focused on the importance of wetlands. The question required students to understand human interference in wetlands. Students also had to have knowledge of the natural state of wetlands and be able to describe utilitarian uses for wetlands, as well as the value of the natural functions of wetlands.

Barrier Islands

Barrier islands function to protect inland areas. They also typically have a brackish area between them and the coast that may serve as a nursery area. Barrier islands shift frequently and may disappear in a large storm as the sand is moved from one side of the island to the other. Humans build resorts and hotels on these sandy beaches for recreation areas, and thus disrupt their natural function. Such areas constantly are undergoing "beach renourishment" projects because they lose sand after major storms. Sand from offshore areas is pumped to the beach to help prevent further damage. Humans also disrupt the natural vegetation and even the nesting sites of sea turtles by settling on these islands.

Lakes

Lakes are freshwater biomes that have a lot of biodiversity. Lakes have a euphotic zone containing plankton and a shoreline area called the littoral zone filled with emergent vegetation. Much of the animal life is also located in this littoral zone, because their food is present there. The bottom is the benthos, which has numerous decomposers and detritovores. Due to the high level of decomposition, the DO is low at the bottom of the lake. In the summer the upper part of the lake, the epilimnion, is typically warmer than the lower part, or hypolimnion. The transition area between the two is the thermocline, an area of water that prevents the two zones from mixing.   In the fall and spring the surface water changes temperature due to the change in season. As a result, the lower water temperature is no longer colder than the surface water. The water will undergo an abrupt "turnover" in which the oxygen laden surface water rapidly moves to the bottom and the lower nutrient-rich water moves to the surface. This turnover is important to maintain productivity in the lake. Unfortunately temperature and oxygen sensitive fish species may die during these turnover events.

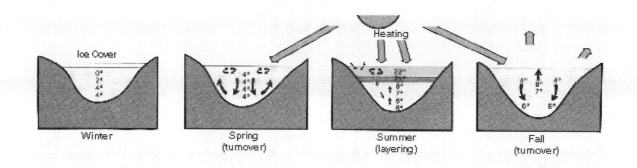

http://www.epa.gov/glnpo/atlas/glat-ch2.html Spring and Fall Turnover

Inland Wetlands

Wetlands are defined as areas that are wet all or part of the year. These areas are frequently used as feeding, breeding, migration, and nesting grounds for a variety of species, particularly waterfowl and shorebirds. Wetlands are recharge zones for aquifers and help filter nutrients and pollutants from surface water. These wetlands have some of the highest NPP on earth. Swamps are wetlands with trees, whereas marshes lack trees and are primarily grass. Bogs are rain-fed marshes and fens are groundwater-fed marshes. Human disturbances include drainage for urbanization and agriculture. The wetlands may also be used for hunting and growing crops such as rice, blueberries, and cranberries.

Chart of Terrestrial Biome Characteristics

The chart lists some of the general characteristics of biomes. The exact flora and fauna is dependent upon which continent the biome is found.

Biome	Precipitation	Temperature	Soils	Representative Flora/Adaptations	Representative Fauna	Major Human Interferences
Temperate Rainforest (cloud forest)	Ample 200-300 cm, very foggy, leaf drip from fog condensing on needles provides much precipitation	Mild winter, cool summers 8-20°C	Old, thin, acidic, nutrient poor	Primarily conifers; Sitka spruce, hemlock Douglas fir, cedar, redwoods, mosses; ferns	Elk; deer; cougar; bears; small mammals; birds, such as northern spotted owl	Logging, dams (salmon)
Tropical Rainforest	> 200 cm, allows decay to occur at a very rapid rate	20-30°C	Old, thin, acidic; nutrient poor because most nutrients are in the biomass; soil may harden to concretelike consistency—laterite soil	Broadleaf evergreen tree species; diverse species spread throughout forest; emergent layer of tall trees with a thick layer of canopy trees below. Trees may have buttresses (expanded bases) due to massive size and shallow roots; epiphytes, such as bromeliads and orchids; lianas (hanging vines); lower levels of forest have smaller bushes, mosses, liverworts if sunlight penetrates; if thick canopy, little to no plant life may be present	Insects; amphibians; reptiles; birds; monkeys; mammals, including sloths; tapirs; Asian elephants; jaguars; tigers	Agriculture. timber harvest, livestock grazing, some mineral mining

Biome	Precipitation	Temperature	Soils	Representative Flora/Adaptations	Representative Fauna	Major Human Interferences
Tropical Seasonal Forest (tropical monsoon; tropical deciduous)	Distinct wet and dry seasons, 150-350 cm	20-30°C	Higher nutrients than tropical rainforest, but still nutrient poor, acidic, and highly leached	Deciduous to decrease water loss during dry season, usually lack buttresses, may have thorns, cacti may be present	Insects; amphibians; reptiles; Australian marsupials, such as koala and kangaroo	Logging, agriculture, invasive exotics, grazing, dams
Temperate Deciduous Forest	75-150 cm, even amounts throughout all seasons	0-20 °C	High nutrients due to deciduous trees, lots of humus and leaf litter	Deciduous trees—oaks, maples, sycamore, beech; Shrubs; ferns; lichens; mosses	Deer, rabbits, squirrels, chipmunks, bears, bobcats, lots of amphibians and reptiles	Logging, Agriculture, tree farms, urbanization
Boreal forest (taiga, northern coniferous forest)	50-150 cm	-5 to 15 °C	Acidic, lots of litter but slow decomposition	Low plant diversity; primarily conifers such as pines, hemlocks, spruce, cedar, fir; some deciduous, like maples, aspen, birch; slow growing season	Wolverines, moose, caribou, bears, elk, migratory birds	Logging, mining, fur trade, dams
Polar Grassland (arctic tundra)	< 10 cm	-20 to 10 °C, Organic material is slow to decompose because so cold, short growing season, low biodiversity	Soils thin; permafrost layer underneath; soils very young—since there is little organic litter; soil is frozen in winter and boggy in summer since water cannot penetrate the permafrost.	Lichens, mosses, sedges, dwarf bushes; plants must reproduce quickly in growing season	Insects, migratory waterfowl, terns, shorebirds, songbirds, lemming, arctic hare, ptarmigan, arctic fox, lynx, weasels, grizzlies, snowy owl, reindeer (Eurasia), musk ox, caribou, mountain sheep, wolves in summer	Too cold for much human activity,
Alpine Tundra on Mountaintops						oil/natural gas drilling and associated transport issues, global warming

Biome	Precipitation	Temperature	Soils	Representative Flora/Adaptations	Representative Fauna	Major Human Interferences
Temperate Grassland	25-100 cm, Fire maintained	-5 to 10 °C	Extremely nutrient rich, lots of humus—partially decomposed organic matter that holds in water and nutrients in soil, arises from grasses dying and decaying in winter	Grasses herbaceous flowering plants (forbs), deep roots to withstand drought, fire, and temperature extremes; root systems may form sod, which prevents soil erosion; in United States, called prairie	Bison, elk, antelope, wolves, coyotes, prairie dogs, black-footed ferrets	Agriculture, livestock grazing if too dry for crops
Tropical Grassland	50-150 cm, prolonged dry season, fire maintained	8-20 °C	Low in minerals, easily leached, may have high levels of Al	Grasses; if savanna, may have sparse tree clumps—acacia trees common; eucalyptus in Australia; adaptations to drought, fire, and heat; deep roots	Migratory grazers – wildebeest, antelope, giraffe, gazelles, zebra, hyenas, vultures, lions, cheetahs	Livestock grazing, agriculture, poaching
Chaparral (Mediterranean; temperate shrubland)	40-60 cm, Fire maintained	0-38° C	Shallow, rocky, nutrient poor	Deep roots; thick bark; Small, leathery, waxy leaves (sclerophyllous); Evergreen; Allelopathy; seeds require burning	Foxes; coyotes; snakes and lizards; mule deer; lots of rodents including chipmunks, jackrabbits, kangaroo rats	Urbanization, fires lead to flooding

Zone	Precipitation	Temperature	Soils	Representative Flora/Adaptations	Representative Fauna	Major Human Interferences
Deserts	< 30 cm	Depends on location, deserts usually at 30° N or 30° S of equator or interior of continents due to rain shadow effect -5 to 30°C		Sparse vegetation, reproduce quickly after rainfall, succulents, extensive roots, allelopathy	Usually nocturnal; may estivate (hibernation when hot); dry feces and highly concentrated urine; large grazers, such as gazelle, oryx; United States—small mammals such as kangaroo rats, coyotes, foxes, snakes, owls, hawks, roadrunners	Off-road vehicles, Overgrazing, Urbanization, oil drilling, mining

Chapter 5 Questions

The following choices are for questions 1-4 and refer to plant adaptations found in the biome.

   a. desert
   b. tundra
   c. boreal forest
   d. temperate forest
   e. tropical rain forest

1. plants adapted for shallow nutrient poor soil; have buttresses
2. plants adapted for lack of water availability in winter; deciduous
3. plants adapted to short growing season; have needles to decrease transpiration in cold climate
4. plants take in CO_2 at night to reduce water loss; have needles; stem photosynthesizes

5. All of the following are adaptations of a desert animal except
   a. copious amounts of urine.   b. nocturnal.
   c. small size.   d. large ears to give off excess heat.
   e. estivate as needed.

6. All of the following will affect how much light can penetrate into aquatic ecosystems except
   a. turbidity.   b. sediment deposition.   c. tannins in the water.
   d. depth.   e. temperature.

7. The biome not found in the southern hemisphere is the
   a. desert.   b. taiga.   c. temperate grassland.   d. temperate rainforest.
   e. tropical rainforest.

8. All of the following are reasons why environmentalists oppose drilling in the Arctic National Wildlife Refuge except the
   a. area has a very short growing season.
   b. permafrost will make it difficult for the area to recover because the soil is so thin.
   c. area is used as a calving ground for caribou.
   d. area has high biodiversity, and it will be difficult to recover all of the species.
   e. area is thought to contain only a moderate amount of oil.

9. Which of the following statements regarding inland wetlands is incorrect?
   a. Marshes are wetlands covered primarily by trees.
   b. Wetlands are frequently used to grow cranberries or rice.
   c. Bogs may be filled with Spahgnum moss.
   d. Wetlands are important as breeding areas for many species of waterfowl.
   e. Wetlands play an important role in the water cycle by allowing infiltration of water.

10. Which of the following statements best describes the strata in a temperate lake in the summer?
a. The surface water layer is much cooler than the deeper layers of water.
b. The large number of decomposers in the benthic zone of the lake increase the oxygen at the bottom.
c. The surface water layer is characterized by a thermocline.
d. The upper stratum of the lake has the greatest amount of DO.
e. The deep water in the center of the lake is warm.

Explanations

1. e. The plants described are found in tropical rain forests.
2. d. The plants described are found in temperate forests.
3. c. The plants described are found in boreal forests.
4. a. The plants described are found in the desert.
5. a. All would be correct except choice *a* because desert animals produce little urine.
6. e. The temperature will not impact the amount of light penetrating the water, whereas the turbidity, amount of sediment, tannins, and depth affect the light penetration.
7. b. The taiga, or boreal forest, is not found in the southern hemisphere.
8. d. One reason the area will have difficulty recovering is that it has low biodiversity and is more easily damaged by disturbance.
9. a. Swamps are covered by trees, whereas in marshes, grasses predominate.
10. d. The upper stratum of the lake has the greatest amount of DO because the euphotic zone is where the producers are located that produce the oxygen. The surface water layer would be much warmer than the deeper layers of water because it is the summer. The decomposers would decrease the dissolved oxygen. The thermocline is in the intermediate strata in a lake. The deep water in the center of the lake would be cold.

Chapter 6 – Population Biology

Key Terms

biotic potential	genetic drift	mortality
carrying capacity	immigration	natality
density dependent factors	island biogeography	overshoots
density independent factors	J curve	population crash
emigration	k-selected species	population density
environmental resistance	life expectancy	r-selected species
exponential growth	life span	S curve
fecundity	logistic growth	stress related disease
fertility	metapopulation	survivorship
founder effect	minimum viable population	

Skills

1. Draw J- and S-shaped curves. Label lag phase, exponential growth, and stabilization phases on each curve. Plot the carrying capacity on the S-shaped curve.
2. Contrast density independent and density dependent controls on population growth.
3. Draw the four types of survivorship curves. Characterize the types of organisms represented by the curves.
4. Compare and contrast r-selected and k-selected species. Give specific examples of each.
5. Summarize the theory of island biogeography. Relate the theory to evolution of species and likelihood of endangerment to isolated species.
6. Predict the effects of small population changes on isolated species.

Exponential Growth

Populations growing without restriction exhibit exponential growth, growth at constant rate per unit time. The resultant J-shaped graph shown below is a graphical representation of a population that increases over time. The mathematical formula is $dN / dt = rN$, where change in the number of individuals (dN) per change in time (dt) equals the growth rate (r) times the number of individuals (N) in the population. The equation is also called the biotic potential of the population, the maximum a population could grow under ideal conditions.

J-Shaped Curve

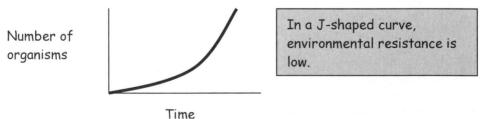

In a J-shaped curve, environmental resistance is low.

Take Note: You must understand the constraints on a population's growth that would cause it to move from exponential growth to logistic growth. You must understand the concept of carrying capacity and know which limiting factors are density dependent and which are density independent.

Carrying Capacity and Logistic Growth

The carrying capacity, K, is the maximum population that can be sustained for an indefinite period by a particular ecosystem. Carrying capacity is not a fixed value, and it may vary from season to season or from year to year. For example, if a drought is present, the carrying capacity will likely be lower than in a year with adequate rainfall. The S-shaped graph generated denotes a population subject to factors that limit its ability to grow. Logistic growth is when population grows exponentially but slows upon reaching K.

$$\frac{dN}{dt} = rN\left(1 - N/K\right)$$

The equation for logistic growth takes into account carrying capacity. The change in the number of individuals (dN) per change in time (dt) equals the exponential growth rate (rN) times (1-N/K). 1-N/K is the relationship between N at any time and the carrying capacity at that time. For example, if N is greater than K, then the population has a negative growth rate. If K is greater than N, then the population has a positive growth rate.

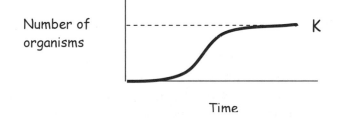

S-shaped, or sigmoid, curve

Sometimes populations exceed carrying capacity, resulting in a population crash. These organisms experience death rates that exceed birth rates, which results in negative growth. Frequently if the population exceeds K, environmental degradation results and the carrying capacity is lowered.

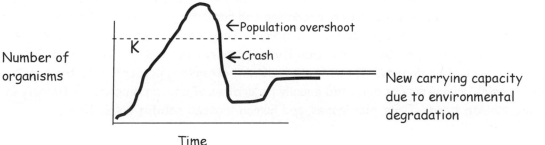

New carrying capacity due to environmental degradation

Rule of 70

Another important number for calculating growth is doubling time. The time for a population to double is (dt) equals 70 divide by the growth rate, or dt = 70/r . For example, a population with a growth rate of 2 percent will double in 35 years (35 = 70/2).

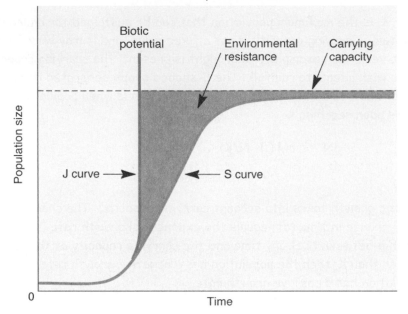

J and S Population Curves

Environmental Resistance and Limiting Factors

Environmental resistance is the sum total of all of the factors that limit population size. Such factors are called limiting factors because they limit population size. As the population increases, so does the environmental resistance. Population density is the population per unit area or volume in aquatic systems. For example, although China and India both have populations over 1 billion, because India is a smaller country in area, it is more densely populated. Density dependent factors are factors more influential as the population density increases. These factors slow population growth by increasing mortality and/or decreasing natality. Density dependent factors impact a greater proportion of the population, not just more individuals in the population. These density dependent factors include; disease; competition; predation and parasitism; territoriality; and increased stress and stress-related diseases, which lead to increased aggression, decreased fertility, decreased immunity, and pathological behaviors. Density independent factors are factors not influenced by changes in population density. These factors do not increase in proportion to the population density, as all organisms are affected equally regardless of population density. Density independent factors include floods, fires, hurricanes, and human-induced habitat disruption.

Spatial Dispersion

Spatial dispersion is how the members of a community are spread in an area. The most common dispersion in method is called clumped dispersion. This pattern is the most common because resources are usually more concentrated in specific areas. Many species remain clumped for protection, for example a zebra herd in the Serengeti. Sometimes the species are clumped because they are a social species, like bees or ants. Some organisms have a uniform dispersion where they are evenly spread out. Examples include most predators due to territoriality and space required for hunting. Any plants that exhibit allelopathy will also exhibit a uniform dispersion pattern. The most unusual distribution in nature is randomly dispersed organisms. This dispersion is rare because resources are usually concentrated. This dispersal pattern can only occur if the environment rarely changes. Examples include the randomly spaced, widely varied tree species in a tropical rain forest or some spider species.

6.5 Spatial Dispersion

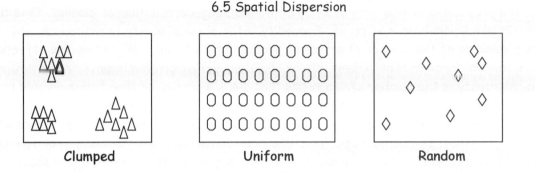

| Clumped | Uniform | Random |

r- and K-Selected Strategists

There are two major groups of organisms based upon their biotic potential. Organisms that can grow rapidly and maintain a high growth rate are known as r–selected (r-adapted) species. Mortality in r-strategists is usually due to density independent factors. Examples of r-strategists include insects, mice, and weeds. K-selected (K-adapted) species are those whose populations are maintained at or near K. Mortality in K-strategists is usually due to density dependent factors. K-strategists are more likely to become endangered, and subsequently extinct, due to the characteristics listed below. Examples of K-strategists include elephants, bald eagles, and elk. The definition of r or K is dependent upon a stable environment. In some cases a species can be either an r- or a K-strategist, depending upon their environment.

r–selected	K-selected
1. Short life	1. Long life
2. Rapid growth	2. Slower growth
3. Early maturity	3. Late maturity
4. Many small offspring	4. Fewer large offspring
5. Little parental care and protection	5. High parental care and protection
6. Little investment in individual offspring	6. High investment in individual offspring
7. Adapted to unstable environment	7. Adapted to stable environment
8. Pioneers, colonizers	8. Later stages of succession
9. Niche generalists	9. Niche specialists
10. Prey	10. Predators
11. Regulated mainly by extrinsic factors	11. Regulated mainly by extrinsic factors
12. Low-trophic level	12. High-trophic level

Take Note: You must be able to explain, with examples, the differences between r- and K-strategists. These life cycles are essential in understanding why species that are threatened or endangered due to human impacts have so much difficulty recovering a viable number of organisms in their populations.

Survivorship

Natality is the production of new individuals by birth, hatching, germination, or cloning. It is the primary way in which populations increase. Fecundity is the physical ability to reproduce, whereas fertility is a measure of the number of offspring produced by a female. Migration also affects population size. Immigration is movement into a population and emigration is movement out of a population. Mortality is the number of organisms that die in a particular time frame divided by the number living at the beginning of the time frame. Survivorship is the proportion of individuals in a population that survive to a particular age. Life expectancy is the probable number of years of survival for an individual of a certain age. Your life expectancy increases for every year you live. Life span is the longest period of life reached by a given organism. The diagram below shows the four types of survivorship curves seen in nature. These are idealized curves. These survivorship curves are generalizations as many organisms may fit different curves at different times of their lives.

Four Basic Types of Survivorship curves for organisms with different life histories

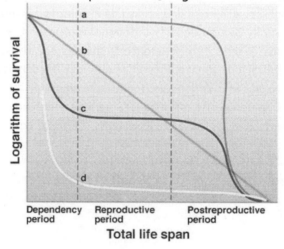

Line a is called a Type I, or late-loss, curve. The likelihood of survival increases with age resulting in more deaths in older organisms. These organisms are frequently top consumers and K-strategists. Line b is a Type II, or constant-loss, curve. The likelihood of death does not change with age, as death is equal in all age groups. Examples include sea gulls, rodents, and some plants. Line c is known as a Type IV curve. These organisms experience early life mortality, which levels off, then rises again later. Such organisms include deer and crabs. They have a large number of highly vulnerable offspring. A Type III, or early-loss curve, is represented by line d. The probability of survival increases as these organisms age. There are many offspring with high mortality, but if they survive, they are likely to live their entire life span. Examples include sea turtles, oysters, and redwood trees.

Take Note: Be able to identify each of the survivorship curves, and be able to give representative organisms of each type of survivorship. You must also know the reason why the organisms have the type of survivorship curve that they possess.

Theory of Island Biogeography

The species equilibrium model, or the theory of island biogeography, was first presented in the 1960s by MacArthur and Wilson. They surmised that the factors that affect diversity on islands would also apply to other ecosystems. These factors include the migration rate to uninhabited area because the faster organisms can migrate, the greater the biodiversity. The extinction rate of already established species in the area is important. If species become extinct slowly the biodiversity will be greater. The size of island also plays a role, because smaller islands are less diverse. The distance from mainland is important, because the organisms may have to traverse a larger space for the new populations to arise. For example, the island of Madagascar, which is very large in size and close to the coast of Africa, has far greater biodiversity than Fernandina, one of the Galapagos Islands hundreds of kilometers off the shore of South America.

Genetic Impact of Population Size

The smaller a population's size, the less the variety in its gene pool, or total available genes in the population. Because the population is so small, one or two small changes in the gene pool, such as the death of a few of the organisms, can result in radical changes to the gene pool. The gradual change in gene frequency due to randomness is known as genetic drift. The founder effect, or demographic bottleneck, occurs when just a few members of a species inhabit a new area. This is an extreme form of genetic drift. They tend to concentrate nocent genes due to inbreeding. The founder effect can be observed in some species during island colonization. The minimum viable population is the minimum number of organisms in a species required to ensure long-term survival of the species.

Metapopulations are populations that have gene flow between geographically separate units. Humans break up natural ecosystems and the habitat decreases in size and quality, with the largest impact on predators and migratory species. This process is known as habitat disruption. Corridors and migration routes are areas that connect fragmented habitats and assist metapopulations in coming together.

Chapter 6 Questions

Use the selections below for questions 1-5.
  a. random distribution
  b. uniform distribution
  c. clumped distribution
  d. density dependent factor
  e. density independent factor

1. territoriality will result in this type of distribution of animals
2. animals that depend upon herds for protection of predators exhibit this type of distribution
3. a cyanide spill from a gold mining operation kills nearly all of the trout in a stream
4. avian influenza kills one half of all of the ducks living in a pond in Southeast Asia
5. the presence of a certain species of cactus has no impact on the presence of other cacti

Questions 6-8 refer to the diagram below.

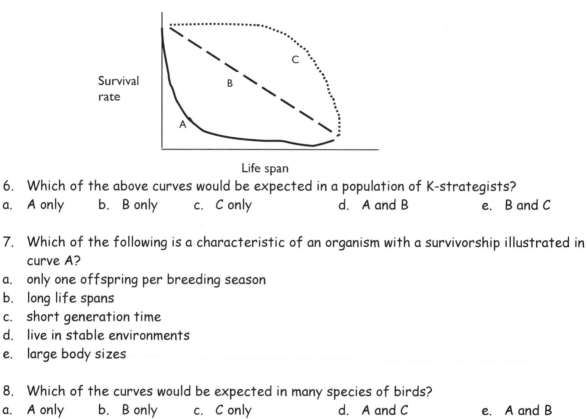

6. Which of the above curves would be expected in a population of K-strategists?
a. A only   b. B only   c. C only   d. A and B   e. B and C

7. Which of the following is a characteristic of an organism with a survivorship illustrated in curve A?
a. only one offspring per breeding season
b. long life spans
c. short generation time
d. live in stable environments
e. large body sizes

8. Which of the curves would be expected in many species of birds?
a. A only   b. B only   c. C only   d. A and C   e. A and B

9. Logistic growth
a. is illustrated by a J-shaped curve.
b. results in a constant rate of increase in the population, unaffected by environmental resistance.
c. maintains a population near the carrying capacity for the population.
d. grows at a constant rate of increase per unit time.
e. occurs when populations are not densely distributed.

10. All of the following are accurate when describing limits on population growth except
a. abiotic impacts are typically density independent.
b. the maximum size a population can attain without inducing environmental degradation is the carrying capacity of the population.
c. zero population growth (ZPG) occurs when the birth rate is greater than the death rate.
d. density dependent factors include competition for mates and resources.
e. if populations overshoot the carrying capacity, the carrying capacity may decrease in a particular ecosystem.

Explanations

1. b. Territoriality results in uniform distribution because it spreads the animals out evenly.
2. c. Herds exhibit clumped distribution because the organisms are trying to avoid predation.
3. e. This is density independent because all animals in the stream are affected equally.
4. d. This is density dependent because a larger proportion of the animals are affected the larger the population.
5. a. If a species is unaffected by the distribution it is random.
6. c. K-strategists have late loss, because they usually have a large size and some level of parenting to protect them during their younger years.
7. c. r-strategists have a short generation time and a survivorship curve as shown in curve A because they have large numbers of vulnerable offspring. They also have large numbers of offspring per breeding, short-life spans, live in unstable environments, and generally have small body sizes.
8. b. Birds tend to have a linear survivorship graph, because they have equal mortality at all age groups.
9. c. Logistic growth maintains a population near the carrying capacity for the population. The curve for logistic growth is shaped like an S, and these populations are affected by environmental resistance. Exponential growth results in growth at a constant rate of increase per unit time. Logistic growth occurs in part because organisms are densely distributed.
10. c. ZPG occurs when the birth rate equals the death rate.

Chapter 7 – Human Populations

Key Terms

birth control
crude birth rate
crude death rate
demographic transition
demography
family planning
infant mortality rate

natural increase
neo-Malthusians
replacement level fertility
total fertility rate
total growth rate
zero population growth

Skills

1. Characterize the four phases of demographic transition, including birth and death rate changes during the transition.
2. Debate the ethics of controlling human population growth.
3. Examine the difficulties in controlling population growth in third world nations.
4. Predict economic and social difficulties that arise during the postindustrial demographic phase due to decreased population growth.
5. Compare age structure diagrams from developed and developing countries. From the diagrams be able to infer information regarding total fertility rates, infant mortality rates, and state of technology and medicine.

Human Population Growth

In 2005 the world population was estimated at 6.4 billion and was growing at a rate of 1.2 percent per year. Using our rule of 70 (70/1.2 = 58.3 years), in 58 years the world population will double to reach 12.8 billion people. The human population has grown relatively slowly until recently, due to the lag time observed in exponential growth. The agricultural revolution allowed more people to be fed, resulting in ample food to support a burgeoning population. In the Middle Ages, populations were held in check by disease, famine, and wars. Major epidemics, including the bubonic plague, influenza, cholera, and typhoid killed large numbers of people in affected areas. The human population began to rapidly increase after 1600, also a time when commerce and communication sprung up between nations. Medical advances and agricultural developments allowed for increases in the population, because most individuals began to survive to reach adulthood. The diagram below illustrates the rapid increase in the human population in the last 500 years.

Date	Population	Doubling Time
5000 B.C.	50 million	?
800 B.C.	100 million	4,200 years
200 B.C.	200 million	600 years
A.D. 1200	400 million	1,400 years
A.D. 1700	800 million	500 years
A.D. 1900	1,600 million	200 years
A.D. 1965	3,200 million	65 years
A.D. 2000	6,100 million	51 years
A.D. 2050 (estimate)	8,920 million	140 years

Source: Population Reference Bureau United Nations Population Division

Thomas Malthus vs. Karl Marx

Thomas Malthus stated that human population growth is not always desirable. He said that the human population can increase faster than food supplies and the inevitable consequences of population growth are famine, disease, and war. Karl Marx felt that population growth was a symptom of poverty, not just a result of poverty. The key to controlling population growth according to Marx was to have social justice for all. Neo-Malthusians believe that humans have reached the earth's K for human population, and we must make birth control our highest priority. Neo-Marxians believe that we must eliminate poverty and oppression to reduce population numbers. Technological optimists believe that humans will be able to raise our K and sustain ourselves regardless of how high our population becomes. We are intelligent enough to find ways to increase our food supply.

Demography

Demography is the study of the size, birth and death rates, and distribution of humans. For example, the highest growth rates in the world are in the Middle East and sub-Saharan Africa, where social mores and politics keep birth rates extremely high. The crude birth is the number of live births/1,000 organisms. The crude death rate is the number of deaths/1,000 organisms. Recall that r is the natural increase in a population. Therefore r = b – d. However, human populations are rarely discreet and we must consider migration into or out of populations, resulting in the total growth rate. With i = immigration and e = emigration, our equation now reads r = (b - d) + (i - e) or population change = (b + i) – (d + e). For example if a population has 20/1,000 births, 10/1,000 deaths, 12/1,000 immigrants, and 4/1,000 emigrants

$$r = (20/1,000 – 10/1,000) + (12/1,000 – 4/1,000)$$
$$r = (0.02 – 0.01) + (0.012 – 0.004)$$
$$r = 0.010 – 0.008 = 0.002 \text{ or } 0.2\% \text{ per year}$$

Replacement level fertility is the number of children a couple must have to replace themselves in the population (usually slightly above two to account for some death). The total fertility rate (TRF) is the number of children born to an average woman during her lifetime. Current fertility rates have been dropping in most parts of the world. For example in Mexico the fertility rate has dropped from 7 to 2.5 in the last 30 years. The infant mortality rate is the number of infants that die prior to one year of age. Zero population growth (ZPG) is the condition when a population is no longer increasing because b + i = d + e. On the whole, birth rates have not increased in the world. Death rates have decreased dramatically in the last century with the advent of technology and medicine.

It must be mentioned that HIV/AIDS is having a dramatic effect on populations throughout the world. It primarily kills young adults, which alters age structure of the population. The life expectancy drops to about 35 years old. The disease decimates the number of available workers and has dramatically increased the number of orphans in many countries. Although the disease skews the age structure, it is not thought that the total world population will be greatly impacted

U.S. Population Growth

The U.S. population is roughly 295 million people. The natural rate of increase in the United States is just 0.6 percent per year, but the total growth rate is high. It is thought that the United States would have a stable population by 2050 without immigration. About 800,000 people legally immigrate to the United States per year and an estimated 500,000 illegal immigrants enter as well. There was a baby boom from the end of World War II through the 1950s due to prosperity and optimism in the populace. In the 1980s there was an echo boom, an increase in reproduction resulting from the large number of baby boomers that reproduced at that time.

Age Structure Diagrams

Age structure diagrams illustrate the proportion of organisms at each age in a population. In human populations, the diagrams represent the percentage of the males and females at each age group from birth to death in the form of a histogram. Humans can be grouped into three main cohorts: prereproductive (0–14 years), reproductive (15–44 years), or postreproductive (45+ years). The shape of the pyramid helps us to determine if population is stable or increasing. A country with ZPG has equal numbers in the cohort groups. A country with a rapidly increasing population has a pyramid shape. If a country has a tapered base, they are experiencing slower growth. You can also infer information about different aspects of the country's development. For example, because a developing country usually has a pyramid shape, you can infer that the populace has less medical care, less technology, and higher infant mortality rates than a developed country with a column

> **Take Note:** You can expect to find age structure diagrams in both the multiple-choice and essay portions of the AP exam. One essay question asked students to evaluate population growth rates from age structure diagrams. They also had to discuss the stages of demographic transition and suggest ways to slow population growth via implementation of government policy. You must understand that an age structure diagram is a snapshot of a country's population numbers *at a given time*. They show the relative percentage of population at each age cohort. Therefore, the individuals from the 0–4 age group do not move up to the 5–9 age group in the diagram. The individuals do not proceed from one age cohort to another in a single diagram. Many inferences can be made from age structure diagrams, but ensure that you do not overextend your interpretation of the data shown. Usually these diagrams illustrate percentage of population at each cohort. You cannot estimate the size of the population as a result of this information.

Age structure graphs for rapidly growing, stable, and declining populations

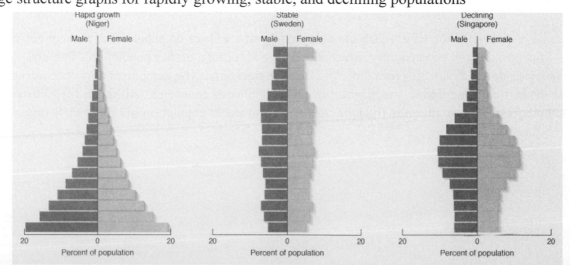

Family Size

Many of the reasons for large families are centered upon a lack of education and wealth in developing countries. For example, if the children are expected to work on the farm or if they are to go to a job each day to help support the family, then more children would be beneficial to a family. Rural families tend to have more children than urban families, because they are frequently needed for labor, and in urban areas, children are far more expensive. If the infant mortality rate is high, more children will be born to compensate for those that die.

The role of women in society plays a huge role in the total fertility rate of a country. The marriage age is extremely important, because the younger a girl is when she gets married, the more likely she is to have greater numbers of children. If girls have numerous educational opportunities, they are more likely to postpone marriage and childbirth until they complete their education. The availability of family planning and birth control measures also controls population growth, because many areas may not have access to adequate assistance in these areas. The availability of pensions is important, because without pensions, children are expected to help their parents when they are too elderly to work.

Other factors that affect family size include religion (more children in Catholic and Muslim families), tradition (customary for the woman to have large numbers of children), and culture (male pride in large numbers of children). Boys are frequently preferred to female children, because they carry on the family name and help assist elderly parents in patriarchal societies.

The lack of women's rights in most areas has a devastating impact on populations. Women are no more than property in many areas where the total fertility rate is exceptionally high. Women must have rights to own property, work, have access to education, and have political rights to lower fertility rates.

In developed countries, there is usually some form of a pension, children are not used regularly for labor, and females have more educational and career opportunities. Children are also a financial burden, as they can be expensive to rear from infancy to adulthood. As a result, the total fertility rate is usually far lower in developed countries.

Graying Populations (Birth Dearth)

Problems can also arise if a population begins to decline rapidly. For example, the large voting base in the older members of the population may alter elections, which could have a negative impact on the smaller, younger populace. Large numbers of older people will have jobs, and it may be difficult to move ahead in the workplace. There will be fewer workers putting funds into pensions and social security, which will result in their collapse without significant restructure. It is expected that there will be labor shortages in menial jobs and a decrease in military strength, because the population has declined so dramatically. Several countries, including Germany, Denmark, and Hungary, are giving incentives to women to encourage them to have more children.

Demographic Transition

During demographic transition, countries move from a preindustrial stage where there are high birth and death rates to a postindustrial stage of low birth and death rates. The idealized model of demographic transition is shown below. During the preindustrial stage, the high birth and death rates are due to high infant mortality rates, malnutrition, and disease due to lack of sanitation and medicine. With the transitional stage comes a decreased death rate due to better medicine and sanitation and an improved standard of living. The birth rate remains high so the country grows rapidly. As industrialization begins, food production increases, health care improves, technology arrives, and the country enters the industrial stage, where the birth rate also begins to decrease and approaches the already low death rate. The postindustrial stage has low birth and death rates so that country reaches ZPG. For demographic transition to occur, four factors must be present: the children must survive to maturity; the standard of living in a country must be improved, along with improved social status for women; and the country must have access to birth control and be educated about its use. Some argue that a demographic trap maintains populations in the transitional phase, which continually increases the population instead of entering the industrial phase, which would begin decreasing birth rates. Social justice may also play a role in transition. The poorer developing countries will need assistance to push into the industrial phase.

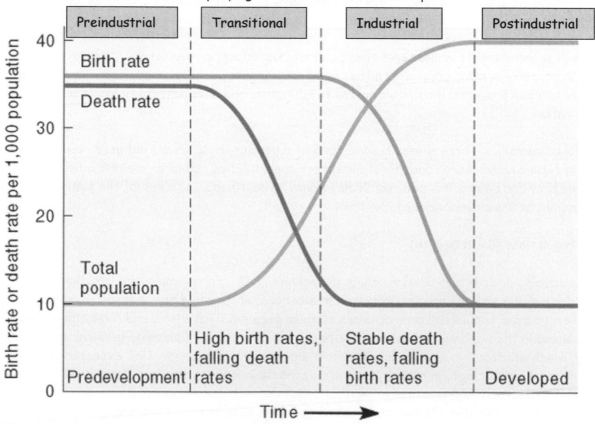

Theoretical birth, death, and population growth rates in a demographic transition accompanying economic and social development

Family Planning and Birth Control

Family planning is used to determine the desired size of the family, and birth control methods are used to help space births to result in healthy infants. Birth control methods include abstinence; avoiding sex during a woman's ovulation; mechanical barriers (diaphragms and condoms); hormones such as progesterone and estrogen (found in The Pill, patches, and Norplant); physical barriers to implantation (IUD); sterilization, and abortion, including RU-486.

Government control of family size has been successful in China. The Chinese became concerned as their population neared 1 billion. As a result, families are limited to one child per family. One reason the program is successful is because China is a communist country, and the populace has little impact on political decisions. The government provides incentives for one child families with increased benefits for the child and family including free education and medical care, preferred housing, and greater pensions. The plan has been successful in lowering the total fertility rate from 6 in 1970 to 1.8 in 1990 and 1.7 in 2005.

There are many drawbacks to a policy limiting population growth. For example, women may be pushed to get abortions and undergo sterilization. In rural areas, the one-child-per-family rule may be ignored because families need the extra helpers on farms. The families are given no freedom of choice, and thus their right to have the number of children they desire is repressed. An additional problem is a disproportionate male to female birth ratio (119-100) because sons are favored over daughters in the Chinese society.

Chapter 7 Questions

1. The population of a country is growing at a rate of 3.5 percent. How long will it take for the population to double?
   a. 10 years   b. 15 years   c. 20 years   d. 25 years   e. 30 years

2. The population of a country is 10 million people. The current growth rate of the country is 1.5 percent. What would be the approximate population in the year 2100 if the country continues to grow at its current rate?
   a. 15 million   b. 20 million   c. 25 million
   d. 30 million   e. 40 million

3. Populations entering the industrial phase of demographic transition exhibit
   a. high birth and death rates.   b. high infant mortality rates.
   c. a lack of medical care.   d. improved status for women.
   e. no access to birth control.

Questions 4 and 5 refer to the following exhibit

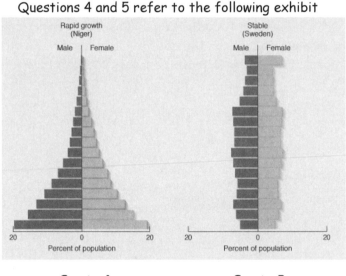

Country A   Country B

4. Country A is likely a
   a. developed country in the postindustrial stage of demographic transition.
   b. developed country in the industrial stage of demographic transition.
   c. developing country in the preindustrial stage of demographic transition.
   d. developing country with low infant mortality rates and low fecundity.
   e. developed country with available technology and medical care.

5. Country B could be all of the following countries except
   a. Hungary.   b. Japan.   c. Germany.   d. Mexico.   e. Sweden.

6. Which of the following is one of the most important issues facing countries in the postindustrial stage of demographic transition?
a. fewer workers for menial entry-level jobs
b. high infant mortality rates
c. poor status of women in the population
d. increased military strength
e. child labor problems in larger cities

Questions 7 and 8 refer to the following statement.
A population of 100,000 frogs experiences 1800 births per year, 600 deaths, 200 immigrants, and 400 emigrants.

7. What would be its growth rate in percent per year?
a. 0.5   b. 1.0   c. 1.5   d. 2.0   e. 2.5

8. What would be its doubling time?
a. 10 years   b. 25 years   c. 35 years   d. 70 years   e. 140 years

9. All of the following are consequences of China's strict policy on one child except
a. more female children than male children.
b. lowered fertility rates in 30 years to below replacement level fertility.
c. less freedom of choice regarding reproduction.
d. increased abortion rates.
e. over 80 percent of the adult females use contraception.

10. For a country to progress into the industrial phase of demographic transition, which of the following changes must be made regarding the status of women in a society?
a. lower the marriage age of women
b. decrease the availability of educational opportunities for girls
c. prevent them from voting and owning property
d. increase the availability of contraceptives
e. lower the availability of abortions in a country

Explanations
1. c. Using the rule of 70, 70/3.5 = 20 years.
2. e. Using the rule of 70, 70/1.5 = 46.67 years. 2,100 is approximately 90 years from now, roughly two doubling times. The population of a country is 10 million people. Therefore the first doubling would produce 20 million, and the second doubling would result in a population of 40 million people.
3. d. Populations entering the industrial phase of demographic transition exhibit declining birth and death rates, low infant mortality rates, access to medical care, and readily available birth control.
4. c. The shape of Country A is indicative of a developing country in the preindustrial stage of demographic transition. Developing countries typically have high infant mortality rates and high fecundity. A developed country would a have a more columnar shape to its age structure diagram and would have access to medical care and technology.
5. d. Mexico is the only developing country. All of the other countries listed are in the post-industrial phase.

6. a. Countries in the postindustrial stage of demographic transition will have difficulty finding laborers in menial entry-level jobs. These countries have low infant mortality rates, high status of women in the population, and decreased military strength due to a lack of young people. Child labor is not typically found in countries in the postindustrial stage of demographic transition.

7. b. Using the formula $r = (b+i) - (d+e)$ 18/100,000 + 200/100,000) - (600/100,000 + 400/100,000) = 20/1,000 - 10/1,000 = 10/1,000 = 0.01 which is 1 percent population growth.
   d. Using the formula $r = (b+i) - (d+e)$ 2,700/100,000 + 300/100,000) – (900/100,000 + 100/100,000) = 30/1,000 – 10/1,000 = 20/1,000 = 0.02 which is 2 percent population growth.

8. d. 70 years. Using the population growth from question 7, 70/1 percent = 70 years.
   e. Using the population growth from question 6, 70/2% = 35 years.

9. a. There are more males than females in the prereproductive cohort because males are preferred in China.

10. d. More contraceptives must be available for a woman to control her reproductive rights. The marriage age should be higher to lower the TFR. Educational opportunities and holding an equal position in the community as men (able to vote and own property) will allow the country to proceed to the industrial phase. By decreasing the availability of abortions in a country, women are sometimes forced into unwanted pregnancies.

Chapter 8 Environmental Health and Toxicology

Key Terms

acute	emergent diseases	risk
allergens	disease	risk assessment
antigens	endocrine disruptors	sick building syndrome
bioaccumulation	environmental health	synergism
biomagnification	morbidity	teratogen
cancer	mortality	toxicology
carcinogen	mutagen	toxin
chronic	neurotoxin	
ecological diseases	pathogens	

Skills

1. Differentiate between chronic and acute exposure to toxins.
2. Characterize the four main ways toxins affect living organisms.
3. Diagram a dose response curve. Relate the terms threshold dose, LD_{50}, and LC_{50}.
4. Identify some major environmental toxins and outline their effects on human health.
5. Relate infectious organisms to their disease and transmission route. Include emergent and ecological diseases.
6. Differentiate between bioaccumulation, biomagnification, and persistence.
7. Evaluate environmental risks currently facing U.S. citizens.

Take Note: Past essays have required students to understand basic epidemiology (who gets infected, where is it prevalent, why does it spread, what is the nature of the disease, when is there an outbreak, and how is it controlled) of prevalent diseases. For example, one essay asked students to link disease incidence to weather changes. Another question asked students to explain disease transmission and the environmental conditions that led to the prevalence of the disease. The question also expected students to understand the role of HIV in increased human mortality rates throughout the world. You must know if diseases are waterborne, respiratory, or vector-borne. Also know how they are transmitted, factors that allow them to persist in populations, and ways to prevent the spread of the disease.

Environmental Health and Disease

Health is defined by the World Health Organization as a state of physical, mental, and social well-being, not just the absence of disease or infirmity. Disease is an abnormal state of the body that disrupts homeostasis. Nontransmissible diseases cannot spread and are frequently the result of poor choices, such as with lung cancer or heart disease. Transmissible, or communicable are able to spread from person to person. These diseases are caused by pathogens, or disease causing agents. Morbidity is illness, and mortality is death. Disability adjusted life years measure disease burden, not just how many people die. Chronic diseases such as heart disease and cancer are no longer present only in developed countries. Figure 8.1 illustrates that much of the disease associated with developed countries will become prevalent in developing countries as infectious disease continues to decline. Transmissible disease is still responsible for one-third of all disease-related mortality. Diarrhea, acute respiratory illness, malaria, measles, and tetanus kill nearly 11 million children under age 5 each year. Pathogens may be transmitted by vectors such as the mosquito. Mosquitoes carry malaria, yellow fever, and dengue. Many diseases are waterborne illnesses, including various diarrheas, hepatitis A, typhoid, and cholera. The pathogens include bacteria (tuberculosis, cholera), viruses (hepatitis, HIV, measles), protozoa (malaria, giardia, trypanosomiasis), and numerous worms and flukes (schistosomiasis, elephantiasis, and guinea worms). For example, giardia is thought to be the largest single cause of diarrhea in the United States. It is spread via fecal/oral transmission and is very common in day cares and nursery schools.

Figure 8.1 Leading causes of global disease burden

RANK	1990	RANK	2020
1	Pneumonia	1	Heart disease
2	Diarrhea	2	Depression
3	Perinatal conditions	3	Traffic accidents
4	Depression	4	Stroke
5	Heart disease	5	Chronic lung disease
6	Stroke	6	Pneumonia
7	Tuberculosis	7	Tuberculosis
8	Measles	8	War
9	Traffic accidents	9	Diarrhea
10	Birth defects	10	HIV/AIDS
11	Chronic lung disease	11	Perinatal conditions
12	Malaria	12	Violence
13	Falls	13	Birth defects
14	Iron anemia	14	Self-inflicted injuries
15	Malnutrition	15	Respiratory cancer

Source: *World Health Organization, 2002.*

Emergent Diseases

Emergent diseases are diseases that were previously unknown or that have not been prevalent for 20 years. Severe Acute Respiratory Disease (SARS) and avian influenza are examples of current emerging diseases. People are concerned about avian influenza. Influenza viruses mutate readily and will pick up genes from other influenza viruses. The great Spanish influenza pandemic of 1918 killed healthy adults, not just the old, sick, and young. It has recently been found to be genetically similar to the avian influenza that many are worried about today. Several viral hemorrhagic fevers, such as Ebola, Marburg, and dengue, continue to have outbreaks throughout the world. West Nile virus is currently present in the United States. It is spread via a mosquito vector and induces encephalitis. It is particularly troublesome because it also infects and kills animals, including numerous bird species. It is difficult to eradicate a disease that has wild animals as a reservoir. A disease that can be transmitted between human and animal populations is called a zoonosis. HIV, the virus that causes AIDS, is also an emergent disease, because it is relatively new. Sixty million people on earth are infected with the virus and 3 million die each year of complications from the disease. AIDS has left 14 million children with one or no parents as a result of the disease. Another example of emerging disease after a period of quiescence is African Sleeping Sickness.

The civil wars in Africa are causing people to move closer to areas previously uninhabited by humans due to the high incidence of African Sleeping Sickness because the insect vector, the tsetse fly, is present. There has been a recrudescence in the disease because people have been forced to move into the less desirable areas.

Ecological Disease

When a disease spreads quickly through animal populations, it is known as an ecological disease. Scientists are concerned about elk and deer wasting disease in North America and its potential spread to humans. The disease is a spongiform encephalopathy (SE) similar to the bovine form also known as mad cow disease. The spongiform encephalopathies are caused by proteinaceous particles called prions that are transmitted from animal to animal. It has been demonstrated that cases of mad cow disease were transmitted to humans, inducing a form of Creutzfeldt-Jakob disease, a human spongiform encephalopathy. Due to the nature of these varying emergent diseases and that many of them are zoonotic illnesses, conservation medicine has developed to allow scientists to link ecological, environmental, and human impacts on disease transmission.

Antibiotic Resistance

There is a resurgence of bacterial disease because drug-resistant organisms are on the rise, as humans travel more and antibiotics are overprescribed and taken incorrectly. Antibiotics are also available in many countries without a prescription. Antibiotic soaps are used frequently and may contribute to disease resistant strains. Some feel the widespread use of antibiotics in livestock and diary industry could also be contributing to the incidence of drug-resistant species.

Toxicology

> **Take Note:** It is of paramount importance that you are familiar with the toxicology of common environmental contaminants. See the chart in the appendices for a detailed listing of numerous pollutants, their sources, environmental effects, human health effects, and ways to prevent release or remediate the presence of the pollutant. Nearly every AP exam thus far has had at least one essay question on pollution and 25–30 percent of the multiple-choice questions are about pollution. You must be able to describe the impacts and origin of the pollutants, as well as be able to give other examples of pollutants and their impacts. For example, at one time there was an essay question regarding mercury toxicity, its origin in the environment, and examples of other heavy metals.

Toxicology is the study of adverse effects of toxins on living organisms. Toxins are chemicals that adversely affect living organisms by disrupting normal metabolic function. All toxins are hazardous but not all hazardous materials are toxins. For example, a material is considered hazardous if it is flammable, explosive, irritating, caustic, or induces allergy.

Figure 8.2 The table below lists the top 20 toxins regulated by the Superfund Act (CERCLA—Comprehensive Environmental Response, Compensation, and Liability Act). They are listed in order of importance in terms of human and environmental health.

1. Arsenic
2. Lead
3. Mercury
4. Vinyl Chloride
5. Polychlorinated Biphenyls (PCBs)
6. Benzene
7. Cadmium
8. Benzo(a)pyrene
9. Polycyclic aromatic hydrocarbons
10. Benzo(b)fluoranthene
11. Chloroform
12. DDT
13. Aroclor 1254
14. Aroclor 1260
15. Trichloroethylene
16. Dibenz(a,h)anthracene
17. Dieldrin
18. Chromium, Hexavalent
19. Chlordane
20. Hexachlorobutadiene

Source: U.S. EPA, 2003.

Effects of Toxins

Toxins sometimes induce allergy, an aberrant immune response to an innocuous antigen. The body will make immune proteins known as antibodies in response to the antigen. When the body is re-exposed to the antigen, the antibodies are released in large numbers. Formaldehyde frequently triggers an allergic response in humans. It is commonly used in the manufacture of furniture, carpeting, and particleboard. Because these items are frequently found in buildings, a building may be labeled as inducing sick building syndrome (SBS). The symptoms may include headaches, sneezing, dry cough, itchy skin, nausea, dizziness, and fatigue. Symptoms typically improve when the individual is away from the building. Other substances associated with SBS include mold spores, carbon monoxide, nitrogen oxides, and cigarette smoke. These buildings typically have inadequate ventilation or are new buildings. Some pollutants weaken the immune system, including pesticides and polychlorinated biphenyls (PCBs).

Many chemicals act as neurotoxins, damaging the neurons in the nervous system. These toxins include the chlorinated hydrocarbons such at DDT, Dieldren, PCBs, and dioxins, the organophosphate and carbamate pesticides; heavy metals such as mercury, and lead; and many industrial solvents. The heavy metals kill nerve cells, and thus the damage is irreparable. The chlorinated hydrocarbons disrupt neuron membrane function. The organophosphate and carbamate pesticides inhibit the enzyme that breaks down the neurotransmitter acetylcholine between neurons and skeletal muscle cells.

Many toxins are endocrine disruptors, including hormone mimics, hormone blockers, and metabolic disrupters. The endocrine system functions in conjunction with the nervous system to maintain homeostasis. Hormones are chemicals produced in one part of the body but that have their effects in another part of the body. Fish have been found that cannot reproduce because many male fish are feminized due to chemical pollutants. Alligators in Lake Apopka, Florida were severely affected by a DDT spill in the 1980s. Ninety percent could not hatch and nearly all the rest of the hatchlings died after birth. The males had small penises and little testosterone. Female alligators had abnormal ovaries and high estrogen. DDT seems to interfere with sex hormones because in humans women seem to get more breast and vaginal cancers and men have low sperm counts and fertility. Also known to be endocrine disruptors are dioxins, PCBs, phthalates in some plastics, several pesticides, and most of the heavy metals.

Mutations are caused by an agent called a mutagen, which induces a change in DNA. Most mutations occur in body cells later in an organism's life. These mutations could result in tumor growth or cancer but most are repaired by cellular repair mechanisms. If mutations occur in gametes, the mutation is passed to the offspring as a birth defect. Cancer is the uncontrolled proliferation of cells that form a tumor. It is the second leading cause of death in the United States. Tumors are induced by carcinogens, which range from radiation and chemicals to viruses. Birth defects are physical; biochemical; or functional abnormalities, such as a cleft palate or spina bifida.

Teratogens (terato—monster; gen—to make) are agents that affect embryonic development, such as drugs or alcohol; radiation; heavy metals; or biological agents, such as Rubella virus. For example, the drug thalidomide was used as a treatment for nausea in pregnant women as it had caused no problems in lab animals. Unfortunately it caused birth defects in humans, primarily affecting limb bud formation, resulting in children born without the upper portions of their limbs.

Nature of Toxins

Toxins are usually either soluble in water or in oil. Water-soluble toxins are more dangerous because water is present virtually everywhere. In the body, the oil-soluble toxins are able to enter cells much more readily than the water-soluble toxins. They also tend to accumulate in fatty tissue in living organisms. The route of exposure of the toxin is important. Toxins can be inhaled, ingested, or absorbed through skin. The inhaled toxins usually are the most dangerous, because we must breathe constantly and our lungs tend to readily absorb toxins due to their structure. Other factors which affect exposure, and thus susceptibility, to toxins are age, genetics and workplace exposure. Some toxins also are not toxic in their ingested form, but are converted by the liver into a toxic chemical.

Bioaccumulation and Biomagnification

Bioaccumulation refers to the buildup of chemicals in the body tissues. Biomagnification is the increase in concentration of a toxin in successive trophic levels, resulting in the concentration of toxins in the predators at the highest trophic level. The higher in the food chain an organism, the more concentrated the toxin will be in its tissues. The chlorinated hydrocarbons are highly persistent and fat soluble, and heavy metals such as lead and mercury are readily biomagnified. Another factor is persistence of the chemical, how long it can remain in the environment. Persistent organic pollutants (POPs) are extremely dangerous due to their ability to be biomagnified. For example, phthalates found in polyvinyl chloride plastic and some deodorants and cosmetics have been shown to be toxic to lab animals. Several of the phthalates are endocrine disruptors. Nearly everyone in the United States has phthalates in their tissues because these chemicals are highly persistent.

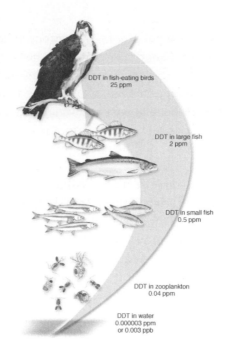

Figure 8.3. Bioaccumulation and biomagnification.

DDT in fish-eating birds
25 ppm

DDT in large fish
2 ppm

DDT in small fish
0.5 ppm

DDT in zooplankton
0.04 ppm

DDT in water
0.000003 ppm
or 0.003 ppb

Chemical Interactions

Chemicals found in the environment can interact in a variety of ways. They may elicit an additive response, where the effect is the sum of individual responses. For example, rats exposed to lead and arsenic show twice the toxicity of an exposure to just one of the toxins. A synergistic response is a response where one substance exacerbates the response to another substance. For example, exposure to asbestos by someone who smokes cigarettes results in a 400-fold increase in cancer rates, not just double the rates as expected. Antagonistic effects are when chemicals negate the effects of another chemical.

Measuring Toxicity

The dose is the amount of a toxin given to a test subject. Typically the higher the dose of the toxin, the more deleterious the effect on the test organisms. Toxicologists expose lab animals to varying doses of a toxin to determine the physiological response to the doses. The resulting graph is called a dose response curve. The threshold level is the first dose at which the effects of the toxin first appear.

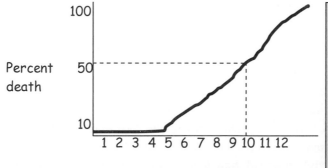

Percent death

100

50

10

1 2 3 4 5 6 7 8 9 10 11 12

Figure 8.4 Dose response curve. The threshold is the first dose in which a response is obvious. There are some natural deaths in the animals at all doses, so the threshold value would be at 5 mg/kg because that is the first dose that the percent death exceeds the natural attrition of the test subjects. The LD_{50} would be 10 mg/kg according to the data given.

Take Note: A dose response curve had to be graphed by the student as a part of an AP essay question. The student had to correlate the graph to the LD_{50} and the threshold dose of the toxin used in the experiment. The question also asked if experimental data conducted on an invertebrate species could and should be feasibly extrapolated to human subjects.

The LD$_{50}$, or median lethal dose, is the dose that kills one-half of test animals or lethal dose for 50 percent in two weeks. The LC$_{50}$ is used for aquatic species who would not be administered a dose of a toxin, but would be subject to lethal concentrations of a toxin in the water surrounding their bodies. LD$_{50}$ are usually given in mg/kg body weight. The lower the LD$_{50}$ value, the greater the toxicity of the chemical. A substance is considered toxic if the LD$_{50}$ is 50 mg or less per kg body weight. Mammals are typically used for these studies, because they are the most similar to humans. Unfortunately results may not be easily extrapolated to humans. Rodents are used, but they may not reflect what happens in human beings, as demonstrated by the thalidomide toxicity in humans. Unrelated species vary in body size, metabolic activity, and physiology. There are ethical concerns with using animals, but our most reliable data is generated using animals in the studies, because computer simulations and cell cultures may not respond as human tissues would to a particular toxin. The amount of time an individual is exposed to toxin is very important. Acute exposures are usually a single occurrence, whereas long-term repeated exposures are deemed chronic. Dose thresholds and LD$_{50}$ studies are important for acute studies of chemicals only, because for a carcinogen, mutagen or teratogen, any amount could create an increased chance of damage in an individual. The effects of the toxin can also be chronic or acute, depending on if the effect is long-lasting or immediate.

The Delaney Clause of 1958 forbade addition of any amount of known carcinogen to foods or drugs. However, that standard was amended in 1996 to state that a substance could be added to food if less than one cancer for every 1 million people exposed over their lifetime resulted from exposure to the chemical.

Risk Assessment

Risk is the probability of suffering harm. A risk assessment estimates the threat of a hazard on human health. The risk must be identified by examining potential and real danger involved, which incorporates a personal perception of risk. Risk acceptability is determining if an activity or exposure to harm is worth the danger of the exposure. For example, scuba diving is perceived by many to be extremely dangerous, but those who participate in the sport find the enjoyment of the activity well worth the perceived harm.

Chapter 8 Questions

1. Teratogens induce abnormalities in which of the following cells?
a. germ   b. embryonic   c. somatic   d. skin   e. brain

2. Which of the following is correct with regard to toxicity of chemicals?
a. Fat soluble toxins are rapidly cleared from the body.
b. Inhaled toxins are readily released from the body because we breathe constantly.
c. Children are more susceptible to inhaled toxins because they breathe quickly.
d. Younger adults are far more susceptible to toxins than middle-aged adults.
e. The lower the LD_{50}, the lower the toxicity of a chemical.

3. Which of the following diseases is not caused by a water-borne pathogen?
a. dysentery   b. typhoid   c. cholera   d. giardiasis   e. tuberculosis

4. All of the following diseases are caused by viruses except
a. AIDS.   b. measles.   c. mumps.   d. schistosomiasis.   e. Ebola.

5. All of the following are characteristics of a hazardous chemical except
a. induces mutation.   b. flammable.   c. causes allergy.
d. irritates to the skin.   e. burns the skin.

6. Which of the following is the best example of an emergent disease?
a. tuberculosis   b. cancer   c. avian influenza
d. cholera   e. giardiasis

7. An endocrine disrupter might have which of the following impacts on the human body?
a. causes difficulty breathing   b. impairs nerve impulse transmission
c. interferes with normal hormonal function   d. induces mutation in body cells
e. causes birth defects

8. The best reason for extrapolating mouse LD_{50} studies to humans would be that mice
a. have a similar size to humans.
b. have a different physiology to humans.
c. are likely metabolize a toxin similar to human metabolism.
d. have different metabolic rate than humans.
e. might have a different response than another species, such as a guinea pig.

9. Which of the following best pertains to risk assessment?
a. People rarely allow their emotions to be involved in evaluating risk.
b. People readily accept the probability of harm.
c. The news media provides an unbiased examination of health hazards.
d. People are mislead by personal experience.
e. People tend to be rational in their concerns about certain hazards.

10. Which of the following is the best example of bioaccumulation?
a. DDT affecting populations of birds of prey
b. lead building up in the bones of children exposed to leaded paints
c. dimethyl mercury contaminating sharks
d. PCBs affecting the reproduction of alligators
e. furans affecting populations of polar bears

Explanations
1. b. Teratogens induce birth defects.
2. c. Fat soluble toxins bioaccumulate in the body. Inhaled toxins quickly enter our bodies and remain due to our lung structure. Older adults tend to be more sensitive to toxins and the lower the median lethal dose, the greater the toxicity of the chemical.
3. e. Tuberculosis is a respiratory disease.
4. d. Schistosomiasis is caused by a blood fluke, a type of worm parasite.
5. a. Chemicals that would induce mutation are mutagens.
6. c. Avian influenza is an example of an emergent disease because it is not currently prevalent. All of the other diseases are, unfortunately, common.
7. c. An endocrine disrupter would interfere with hormone function. A neurotoxin could cause difficulty breathing and impair nerve impulse transmission. A mutagen would induce mutation in body cells, and a teratogen would cause birth defects.
8. c. The best reason for extrapolating mouse LD_{50} studies to humans would be that mice likely metabolize a toxin similar to human metabolism. The fact that mice have a different size, physiology, and metabolic rate would make the case for not extrapolating human data and because they might respond differently from a guinea pig, they likely would not respond the same as a human.
9. d. People allow their emotions to be involved in evaluating risk and they have difficulty accepting the probability of harm. The news media does provide a bias examination of health hazards and people can be irrational about their concerns regarding certain hazards.
10. b. All of the other examples are of biomagnifications of toxins in apex predators.

Chapter 9 – Food and Agriculture

Key Terms

agribusiness	humus	soil
anemia	kwashiorkor	soil horizons
aquaculture	malnourished	soil profile
contour plowing	marasmus	subsoil
cover crops	mulch	sustainable agriculture
desertification	mycorrizae	strip farming
erosion	obese	terracing
famine	overnourished	topsoil
food security	perennial species	undernourished
genetically modified	reduced tillage	waterlogging
organisms	regolith	
green revolution	salinization	

Skills

1. Compare and contrast the terms malnourished, undernourished, and overnourished.
2. Identify and describe some chronic diseases associated with malnourishment or undernourishment.
3. Relate famine to social, environmental, and economic forces.
4. Identify major plants used for food in the world. Contrast subsistence agriculture with agribusiness.
5. Identify major soil characteristics. Diagram a soil profile.
6. Identify and describe soil tests that would assist in maintaining soil fertility and facilitate agriculture.
7. Differentiate between the soil problems of erosion, salinization, waterlogging, and nutrient depletion.
8. Identify and describe agricultural practices that would ameliorate erosion.
9. Debate the costs and benefits of genetically modified organisms.

Undernourished Versus Malnourished

Many people throughout the world do not have adequate food or specific nutrients for survival. Undernourished means that an individual is not obtaining the number of calories per day, approximately 2,200, which a body requires. Malnourished means that an individual is lacking a specific nutrient. For example, individuals who have a diet of primarily starch (from corn, rice, or manioc), have a vitamin A deficiency. Vitamin A deficiency is the leading cause of preventable blindness, because eyes require vitamin A to properly function. Anemia is the result of dietary deficiency of iron. Anemia affects childhood development and increases the risk of maternal deaths in childbirth. Iodine is required for normal thyroid function to maintain metabolism. Chronic iodine deficiency can result in inflamed thyroid (goiter), stunted growth, and reduced mental capacity. Folic acid deficiencies are linked to neural tube defects, such as spina bifida and anencephaly. The two major protein deficiencies are kwashiorkor and marasmus. Kwashiorkor is a deficiency of just protein, whereas marasmus is a disease of undernutrition and malnutrition of protein. Children with these diseases suffer from stunted growth, mental retardation, and weak immune systems. Overnourishment is typically seen in developed countries. Individuals in these countries ingest foods high in salt, fat, and sugar, which tend to lead to obesity.

Food Security

Food security is the ability to obtain food on a daily basis. Unfortunately most individuals who are undernourished are also extremely poor. Poverty contributes greatly to the lack of food security, because impoverished people are less able to have food reserves and lack the ability to grow their own food. A severe food shortage is called a famine, and famines are characterized by food shortages, starvation, social upheaval, and economic turmoil. Many people migrate from impoverished famine afflicted areas to refugee camps in the hope of finding food and work.

Take Note: One essay question asked students to compare the production of grains to the production of meat for human food resources. Students were expected to know human nutritional requirements and be able to explain environmental impacts of raising animals for meat.

Types of Food

The major food crops in the world are from three plants: corn, rice, and wheat. In some areas, potatoes, barley, oats, and rye are also important. Fruits and vegetables make up very little of the human diet overall, even though they are high in fiber, vitamins, minerals, and carbohydrates. Sixty percent of all meat is raised in developing countries, but they consume only one-fifth of commercial animal products. In the United States, livestock is typically raised in a feedlot, where they can be quickly brought to market weight with a diet designed to add body mass. Feedlots are associated with high levels of air and water pollution, including nitrates and fecal bacteria from animal feces entering water and hydrogen sulfide and particulates entering the atmosphere. Livestock wastes are frequently stored in large lagoons, which may be breached during adverse weather conditions. The crowded conditions of a feedlot result in the animals having frequent infections, thus resulting in antibiotic use and the inferable emergence of antibiotic resistant bacteria. The odor produced in feedlots can be extremely unpleasant. One-third of grain raised in the world goes toward livestock production. Many feel that we could feed far more humans if we avoided feeding grains to livestock, but the Food and Agriculture Organization of the UN claims that the grains raised to feed livestock would not be produced if there was no demand for the grains.

Seafood

Seafood accounts for approximately 15 percent of all animal protein ingested by humans. Unfortunately due to overharvesting and habitat destruction, the world's fisheries are in desperate need of regulation and maintenance. There are several different methods used to capture fish and shellfish. Funnel-shaped trawl nets are used to harvest shrimp and other benthic species from the ocean floor. Trawl nets can also be used to harvest fish species by dragging the net through the water and forcing the fish to the end of the net. Long-lining is a long fishing line to which several shorter lines with hooks are attached. The lines are strung out, sometimes over miles. The crew reels the line back in and pulls off the fish on the hooks. Gill nets are long nets strung out vertically in the water. The fish cannot see the nets, and they swim through them, getting caught in the net around their gills.

One of the greatest impacts of these fishing methods is bycatch, the nontarget species inadvertently caught while employing these various fishing methods. These species may be kept if they have commercial value, but frequently they are undesirable or even protected species. Dragging trawl nets frequently kills the bycatch, so many kilograms of dead organisms are returned to the ocean for each kilogram of marketable food. For example, endangered sea turtles frequently drown when caught by a shrimp trawler; these deaths can be prevented by attaching a turtle exclusion device (TED) to the front portion of the net, which allows the turtle to swim to safety. Additionally the United States Fish and Wildlife Services (USFWS) says that long-lining is one of the greatest detriments to seabird populations.

Aquaculture is growing aquatic species for consumption. Aquaculture dramatically decreases deaths of nontarget species. Fish, including tilapia and catfish, can be reared relatively easily in inland ponds. Coastal fish farming can destroy estuaries and mangrove swamps as the native vegetation is removed to establish the farms. Crustaceans, such as shrimp and lobsters, and mollusks, such as oysters and clams, can be raised on farms. Fish ranching is a similar idea, but the fish are raised to a size where they can survive easily in the wild and then released. Commonly ranched species include salmon and trout. Some red and brown algae are even cultured for harvest.

Farm Policy

Many countries provide subsidies for farmers to assist in food production. In the United States, crops such as corn, wheat, cotton, rice, and peanuts are heavily subsidized. Subsidies cause an increase in surpluses and farmers can sell the crops cheaper than they cost to produce. They sell to undeveloped countries because the price of local food is more expensive than the subsidized U.S. crops. In 2005, the World Trade Organization ruled American subsidies are illegal because they were causing food prices to be distorted.

Green Revolution and GMOs

The green revolution refers to the development of high-yield crop plants. These plants were derived by breeding high-yield plants to one another through years of careful cultivation. These plants require optimum levels of fertilizer, water, and pest protection to produce large yields. A faster way to alter plant genetics is via genetic engineering. Genetic engineering is taking a gene from one organism and putting it into another organism, now called a transgenic organism or genetically modified organism (GMO). The USDA has concluded that transgenic foods are as safe as unaltered foods, and therefore they require no labeling in the United States. Many of the vegetables and grains consumed in the United States are transgenic organisms. It is likely that 60 percent of U.S. food products contain genetically modified (GM) food crops. Benefits include potentially higher yields, genes from any species may be used, and toxins might be removed. Research indicates that these crops do not persist in the wild and that it is highly unlikely for the genes to enter wild strains of the crops. One of the most beneficial modifications has been creating crops with resistance to stress or herbicides. For example, several grains have been modified with a gene that resists the common herbicide Roundup®, produced by Monsanto. The crops can then be sprayed with Roundup® to kill weeds, but the crops will persist without impact. The major benefit is a decrease in erosion because conservation tillage methods are more easily used. Another helpful GMO is golden rice, which contains the gene for a precursor of Vitamin A. Vitamin A has been extremely beneficial in developing countries that have had problems with blindness due to Vitamin A deficiencies.

Some plants have been modified to secrete their own pesticide by inserting a gene from the bacterium *Bacillus thuringiensis* (Bt) that destroys caterpillar and beetle larvae into the crop. As a result of using Bt, farmers do not have to use as many pesticides. Goats have been genetically modified to secrete various products in their milk. Animals are also modified to grow more quickly on less food.

Many feel that GMOs are dangerous, even calling them "Frankenfoods." Some consumers feel they are unsafe, because we do not know the effects the modified plants may have on each other and the environment. For example, the adverse effects on nontarget species are a concern. People are concerned about the Bt toxin in the pollen of corn might potentially damage monarch butterfly larvae when the pollen spreads outside the field. Some are concerned about patenting living organisms and the ethics of such a practice. Others are concerned about the economic aspects of a company creating a GMO upon which only their product works. Also, there is a concern that a transgenic organism might induce allergy in people allergic to the product containing the original gene. Genes cannot be made. The gene must already exist in another organism to be used.

> **Take Note:** Agriculture issues are important in environmental science, and GMOs are extremely controversial. Ensure that you can provide arguments both for and against the use of GMOs in food crops.

Soil

Soil is the weathered portion of the earth's crust that can sustain life. Young soils are not leached of their nutrients. Older soils are leached and have little organic material remaining. The parent material of soil is rock broken down by chemical and physical weathering. Humus is the dark-colored organic material that remains after decomposition of leaf litter, droppings, and plant and animal remains. Leaching is when minerals or matter is dissolved in water percolating downward, thus making the nutrients unavailable to plants. The zone of illuviation is the area in the deeper levels of the soil where the leached matter is deposited. Illuvial material includes iron; humus; and clay, depending on the soil type.

Soil Organisms

Soil organisms include bacteria and fungi, which function as decomposers. Algae can be present on the surface of the soil. Both round worms (nematodes) and segmented worms may assist in aeration of the soil. Insects (particularly ants), roots, snakes, gophers, groundhogs and moles all tunnel in soil and aerate it.

Chemical Properties of Soil

The pH of the soil should be taken to determine if the soil is too acidic or basic. Soil pH should range between 6 and 7 for most plants. At low pH, aluminum, iron, boron, and manganese are more soluble and more available to plants. Aluminum can be toxic to plants when extremely available. Potassium, iron, and manganese are less available in very alkaline soil. To make the soil more acidic, you can add ground sulfur or aluminum sulfate. To make the soil more alkaline, lime can be added. The nutrient levels of the soil must also be determined. The elements that most plants require are nitrogen, potassium, phosphorus, calcium, magnesium, and sulfur. Nitrogen and phosphorus are frequently limited factors, and testing for nitrate and phosphate in the soil will determine if a fertilizer must be applied. Other micronutrients such as iron, manganese, magnesium, and selenium should be measured to determine if they are a limiting factor in the area. The amount of humus can be measured as well and will let you know the amount of organic material in the soil. Humus is important for its nutrient benefits, water holding capacity, aeration capacity, allowing root growth, and increasing porosity of the soil. The salinity of the soil may also be an issue, particularly in irrigated areas.

Physical Properties of Soil

The soil color can be very important. Color can be indicative of the soil's nutrients. For example, dark brown or black soil has lots of humus and therefore, a high organic component. Red soils may contain a lot of iron. The pore spaces in between the particles of soil create soil property of porosity. Very porous soils hold more water and air in the pore between the soil particles. The porosity of the soil influences the permeability of the soil, or the rate at which water and air move through the soil. The more porous a soil, the lower the permeability. The texture of the soil is determined by the relative amounts of different-sized inorganic particles. The three particle sizes of soil are sand, 0.05–2 mm; silt, 0.002–0.05 mm; and clay, < 0.002 mm. Sand has high permeability because of low porosity. The particles are so large that the spaces between them are large as well, resulting in good drainage and aeration. Clay has low permeability because it has high porosity. The particles are tiny, which allows smaller pores around them, thus holding water and preventing water permeating through the soil. The soil mixture called loam has approximately equal portions of each texture type and is considered a soil ideal for agriculture.

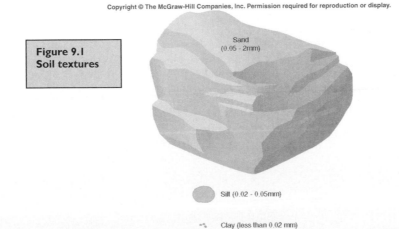

Figure 9.1 Soil textures

The soil texture pyramid shown below will allow you to determine the texture of the soil after you determine the relative amounts of sand, silt, and clay in the soil. For example, if your soil was found to have 50 percent sand, 20 percent silt and 30 percent clay, you use the pyramid to determine the texture. First, look along the bottom of the pyramid until you find 50 percent. Then look along the right side finding 20 percent silt. The left side of the pyramid is for the 30 percent clay. Look for the intersection of the three lines. In this example, the soil type would be sandy clay loam.

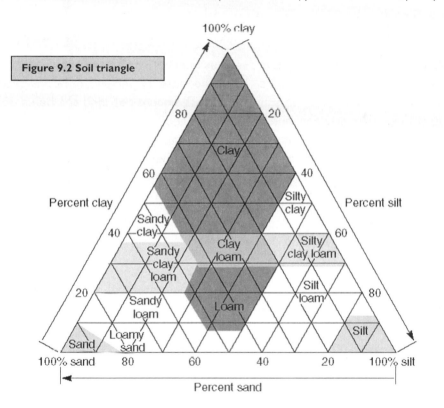

Figure 9.2 Soil triangle

Soil Horizons

Soil horizons are stratified layers in soil. To see the horizons, a soil profile is taken to classify the soil. The O horizon is the organic material on the surface of the soil. The plant litter accumulates and decays. The A horizon, or topsoil, is below the O horizon. It is frequently rich with accumulated humus. The E horizon (eluviated) is a heavily leached soil area that sometimes develops between the A and B horizon. The B horizon or subsoil, is typically a zone of illuviation for the leached material from the A and E horizons. It may be rich in humus, clay or iron. The C horizon, or regolith, contains weathered rock and sits upon the parent material. The color, pH, and depth of the horizons will assist you in identifying soils. The soil types are classified into twelve soil orders.

Major Soil Orders

Spodosols form under coniferous forests. With their layer of acidic pine litter, they have a whiteish, ashy, leached E horizon. This soil is not suited for agriculture due to the acidity and nutrient leaching. Alfisols are formed under temperate deciduous forests. The topsoil in an acfisol is usually gray brown or brown in color, indicative of high levels of nutrients. The relatively high amount of precipitation leaches the A and E horizons, but the soil fertility is maintained by the constantly replaced litter.

Mollisols are found in temperate grasslands. They are fertile with thick, dark brown topsoil layer indicating high levels of humus, and they are subject to little leaching due to dry seasons. Aridosols form in arid regions. There is little leaching because of little precipitation but there is also not much organic matter to replace nutrients. Histosols are formed in waterlogged areas and result in incompletely decayed organic material. Oxisols and ultisols form in hot, wet areas. They tend to be low in nutrients because they are highly leached due to the large amounts of rainfall and lack of litter.

Soil Problems

Less agricultural land is being used now than in previous generations because we have better crop varieties, fertilizers, irrigation, and pesticides. Land productivity per unit acre has greatly increased over time. Unfortunately at the same time, vast amounts of land are being degraded by poor agricultural practices, urbanization, and pollutants. Land is considered degraded if the soil lacks fertility or is eroded; if the surrounding water is contaminated or has more runoff than is typical; if the amount of vegetation declines; if the area has decreased NPP; and if there is less animal biodiversity.

Erosion

Erosion can be due to water, wind, or ice. Erosion is the movement of soil from one place (usually the desired location) to another (usually undesirable location). By decreasing the amount of topsoil, plant growth is limited due to the diminished soil fertility. The total amount of soil lost from cropland is thought to be 25 billion metric tons. There are three types of water erosion. Sheet erosion water moves down a slope and erodes topsoil evenly. Rill erosion cuts shallow channels in soil. Gully erosion is the most severe as it cuts deep channels in soil. It usually occurs on deep slopes. Streambank erosion is the loss of the sides of a stream as the water flows along the soil. Streambank erosion usually arises from cattle grazing in riparian areas or removal of vegetation from riverbanks. Wind erosion occurs most severely in areas with dry seasons and flat terrain. For example, African dust can be detected in St. Petersburg, Florida, due to summer dust storms in the Sahara desert. Soil erosion also leads to increased sediment flowing into surface water, which decreases water quality by increasing turbidity. If the soil is contaminated with fertilizer or pesticides the erosion may pollute the water and induce eutrophication or animal death from the pesticide.

Nutrient Depletion

When crops are harvested nutrients are removed. Farmers replenish these nutrients with fertilizers. These fertilizers are high in phosphates and nitrates, which may induce eutrophication in surface water. Nitrates in groundwater have also been shown to be a problem, as high levels of nitrates in water may be fatal to infants.

Salinization and Waterlogging

Salinization occurs when land is irrigated over a long period, especially from wells. Salt collects in topsoil when the water evaporates and stunts plant growth, which reduces crop yields. The excess salt could be flushed by precipitation. To decrease salinization, irrigation can be via a mechanism that does not allow much evaporation, such as underground pipes or drip irrigation.

Another option in these salt damaged areas is to plant salt tolerant crops, such as barley.

Another problem is waterlogging, which arises when soil is saturated. The plant roots die due to the lack of oxygen required for cellular respiration.

Desertification

Desertification is the conversion of marginal lands to desert due to climate and human behavior. Rangeland and pastures are the most likely to become desertified, as these areas are too arid for crop cultivation. The likelihood of desertification is increased by overgrazing and deforestation. To fight desertification, vegetation is planted to reduce erosion and increase water holding capacity.

Soil Conservation and Regeneration

Sustainable agriculture, or regenerative farming, seeks to produce food and fibers on a sustainable basis while repairing the damage done to farmlands by earlier degrading practices. Soil can be managed to remain productive for a long period. There are several methods of cultivating crops that will decrease water and wind erosion. The major theme to all of them is to avoid conventional tillage, plowing to turn soil to plant seeds. Conservation tillage involves leaving the root residues from prior crops in the soil to hold it in place to prevent erosion. One example of conservation tillage is no tillage, where soil is not turned over by a plow before planting. The seeds are planted in a furrow. This method not only decreases erosion because the soil is not broken up, it also increases the soil's ability to hold water due to an increased amount of organic matter in the soil because it creates a level of humus from the decaying crop remains. The biggest problem is that pests and weeds may persist in the area, so more pesticides and herbicides may be required to maintain the crops.

Crop rotation is growing different crops in the same field over time. The soil constantly has roots embedded in it, which decreases the erosion. The extra benefits of crop rotation include fewer pests/diseases, because pests/diseases are crop specific, and using legumes, which allow nitrogen fixation to occur, thus increasing the amount of nitrates available to subsequent crops. Contour plowing is planting fields to conform to the natural topography such as running rows around a hill, not up and down the hill. Erosion is less likely if the water cannot run down rows. Strip cropping is alternating strips of different crops grown at the same time but harvested at different times, which will hold in the soil, particularly if planted along the land contours. For example, several rows of corn may be grown between swaths of wheat. Terracing creates flat areas in mountainous terrain, which decreases the likelihood of erosion. Shelterbelts are rows of trees that prevent wind erosion. They may be harvested or simply present to prevent wind from moving the soil. Alley cropping or agroforestry is planting strips of trees between crops. Another option is planting cover crops, such as rye, alfalfa, or clover planted after harvest to protect soil from erosion. Soil may be permitted to lie fallow and allow native plants to hold the soil in place. Mulch may be placed over the soil, which will not only prevent the soil from being displaced, it can also decompose to provide humus to the area. A farmer should also use land classification to identify easily erodible land that should not be used for agriculture. The National Resources Conservation Service of the Department of Agriculture (originally the Soil Conservation Service established by the Soil Conservation Act of 1935 to address erosion control) has classified soils based upon erodibility and can assist in developing a management method for the area.

- 90 -

Preserving Fertility

To increase absorption of water and minerals, spores of basidiomycota fungi (mushrooms and puffballs) may be added to soil, which will form mycorrhizal relationships with plants to increase absorption. Implementing crop rotation with a legume crop will increase the nitrogen level of the soil. Organic fertilizers such as manure, crop residue (green manure), bone meal, and compost are recommended. The exact composition varies but these fertilizers are slow acting and long lasting. The greatest drawback is that the exact composition is not known. Natural fertilizers have the additional benefit of increasing the water-holding capacity of soil. Inorganic fertilizers are manufactured from chemical compounds and have a known composition. They are readily available and easily applied. They quickly become available to the plant and rapidly increase the soil fertility. These inorganic fertilizers have several drawbacks. They can leach away into groundwater and enter surface runoff. They cannot increase the water-holding capacity like organic fertilizer. They release the greenhouse gas N_2O and tend to reduce the oxygen content of soil.

Chapter 9 Questions

Use the following choices for questions 1-4.

        a. aridosols
        b. alfisols
        c. mollisols
        d. spodosols
        e. entisols

1. found in most of Canada
2. found in prairies
3. found in the Mojave desert
4. found in Tennessee and North Carolina

Use the following choices for questions 5-8.

        a. O horizon
        b. E horizon
        c. C horizon
        d. A horizon
        e. B horizon

5. contains a zone of illuviation
6. easily leached in temperate deciduous forests
7. contains the weathered parent material
8. topsoil

9. Using the soil triangle (fig. 9.2), determine the type of soil that would contain 30 percent sand, 40 percent silt, and 30 percent clay.
a. sandy clay loam   b. silty clay loam   c. loam   d. clay loam e. silt loam

10. Which of the following erosion prevention methods would be appropriate for an area with steep slopes?
a. strip cropping   b. alley cropping   c. terracing   d. agroforestry   e. mulching

11. What type of soil has the greatest porosity?
a. sand   b. silt   c. clay   d. loam   e. sandy loam

12. The leading cause of preventable blindness in the world is a deficiency in
a. iron.   b. calcium.   c. vitamin A. d. vitamin D. e. iodine.

13. Which of the following types of fishing techniques would be used to harvest shrimp?
a. drift net   b. trawl bag   c. long-lining   d. purse seine net   e. hook and line

14. Which of the following impacts might occur on irrigated arid land?
a. salinization   b. waterlogging   c. nutrient depletion   d. compaction   e. laterization

15. Which of the following is a chemical property of soil?
a. color   b. texture   c. porosity   d. pH   e. permeability

Explanations

1. d. Spodosols are found in Canada. They form under boreal forests.
2. c. Mollisols form under temperate grasslands and thus would be found in prairies.
3. a. Aridsols would be found in a desert.
4. b. Alfisols would form under the temperate deciduous forests found in Tennessee and North Carolina.
5. e. The B horizon contains a zone of illuviation.
6. b. The E horizon is easily leached in temperate deciduous forests.
7. c. The C horizon contains the weathered parent material.
8. d. The A horizon is also known as topsoil.
9. a. Sandy clay loam because the lines intersect at the lower right portion of the clay loam texture range.
10. c. Terracing (cutting terraces into the steep slopes) will decrease erosion.
11. c. Clay has the greatest pore space because they have the smallest particle sizes.
12. c. Vitamin A is necessary for normal eye function, thus a deficiency would cause blindness. A deficiency in iron causes anemia, and a deficiency in calcium or vitamin D will cause rickets. An iodine deficiency interferes with the function of the thyroid.
13. b. A trawl bag would be used to harvest shrimp because they are small. Drift nets and purse seine nets are used for larger aquatic organisms, and long-lining and hook and line fishing would not work since the hooks would be too large for such a small species.
14. a. Salinization might occur on irrigated land.
15. d. pH is the only chemical property listed. All other answers are physical properties.

Chapter 10 – Pest Control

Key Terms

biocide	fumigants	natural organic pesticides
biological controls	fungicide	organophosphates
biological pests	herbicide	persistent organic
carbamates	inorganic pesticides	pesticide
chlorinated hydrocarbons	insecticide	pesticide treadmill
circle of poison	integrated pest	pest resurgence
DDT	management	pollutants

Skills

1. Identify the major types of pesticides and the pests they are designed to control.
2. Correlate biomagnification and bioaccumulation of DDT in raptors to their endangerment.
3. Summarize the history of pesticide use in the world.
4. Argue the benefits of pesticide use versus the damage done by pesticides to the environment.
5. Characterize integrated pest management. Distinguish between the various methods of IPM.
6. Discuss the regulation of pesticides.

Pesticides and Pests

Pesticides are chemical substances that kill organisms. Pesticides include herbicides which kill plants; rodenticides, which kill rodents; fungicides, which kill fungi; nematocides, which kills nematodes; and insecticides, which kill insects. Pesticides that kill many species are known as biocides. The United States uses about 5.3 billion pounds of pesticides per year. Nearly half of that amount is the chlorine and hypochlorites used to treat water to prevent waterborne illness such as dysentery, cholera, giardiasis, and cryptosporidiosis. Conventional pesticides make up the other half. Cotton is the crop that has the highest rate of insecticide application. Golf courses frequently have higher pesticide application rates than farms. The United States uses 20 percent of the world's consumption of conventional pesticides, about 1.24 billion pound per year.

History

Ancient Romans burned infested locust fields to prevent their spread. The Sumarians used sulfur as an insecticide. In China mercury and arsenic were used on ectoparasites and added ants to orchards to kill caterpillars. Spices were valued because they decreased pests in foods. The modern era of pesticides began with the discovery of the insecticidal properties of dichlorodiphenyltrichloroethane (DDT). The pesticide was inexpensive to manufacture, stable, easily applied, and lethal to insects. Mass production began during WWII. The pesticide helped eradicate malaria in the United States by destroying the *Anopheles* mosquito, the insect vector for the protozoal disease. DDT is a broad spectrum pesticide that kills a wide variety of insects. Narrow spectrum pesticides can only control a few pests. An unanticipated effect of DDT in bird populations was not immediately recognized. There was a huge decline in the numbers of some bird species, including peregrines, pelicans, cormorants, bald eagles, and osprey. The birds were bioaccumulating large amounts of DDT, which was undergoing biomagnification because these birds are at the top of the food chain. The effect that DDT has on bird species is impaired reproduction because of fragile eggshells. In 1962 Rachel Carson published *Silent Spring*, which pointed out the adverse impacts of DDT use. DDT was banned in the United States in 1972.

Chemical Groups of Pesticides

The inorganic pesticides include simple compounds of sulfur, arsenic, copper, lead, and mercury. These chemicals are toxic in addition to being environmentally persistent. Natural organic pesticides are usually extracted from plants. For example, tobacco produces toxic nicotine sulfate, which is toxic to insects and mammals. Pyrethrum is extracted from chrysanthemums and is still used in animal dips and flea shampoos. Rotenone, used to kill fish, is derived from root of derris plant. The fumigant chemicals include carbon tetrachloride, ethylene dichloride, and methyl bromide, all of which easily volatilize and penetrate materials. They are used to sterilize soil and decrease infestation in stored grains.

The chlorinated hydrocarbons (organochlorines) are a toxic group of pesticides, and nearly all are banned or drastically restricted in United States. They include DDT, aldrin, kepone, dieldrine, chlordane, and toxaphene. They are nerve toxins that acutely cause nausea, vomiting, convulsions, and death by respiratory failure by interfering in the transmission of nerve impulses. They also are linked to all fertility disorders, and the Environmental Protection Agency (EPA) classifies them as probable human carcinogens. They are highly persistent in the environment, are fat soluble, and subject to biomagnification. The organophosphates include malathion, parathion, and tetraethylpyrophosphate. They were used in World War II as nerve agents because they inhibit the enzyme that breaks down the neurotransmitter acetylcholine released at the neuromuscular junction. They break down relatively quickly and are not likely to bioaccumulate as the organochlorines. However, they are more expensive to produce and more toxic in lower amounts than the chlorinated hydrocarbons because they are rapidly absorbed.

Subacute doses induce headache, slow heart beat, confusion, vomiting, and difficulty breathing. Acute doses may lead to paralysis, tremors, coma, and death. An organophosphate that does not affect the nervous system is called glyphosate, commercially sold as Roundup®. Carbamates are used as insecticides, herbicides, and fungicides. They have the same mode of action, toxicity, and lack of persistence and bioaccumulation found in the organophosphates—examples include carbaryl (sevin) and aldicarb (Temik). Carbamates are toxic to bees, and caution must be exercised if bees pollinate crops. Many of the pesticides have chronic health effects that include cancers, birth defects, immunological problems, neurological problems, and endometriosis.

Pesticide Benefits

The use of pesticides has eradicated many insect vectors, including mosquitoes (malaria, dengue, yellow fever), rat fleas (bubonic plague), typhus (lice and fleas), and tsetse flies (African sleeping sickness). Many countries feel that the health risks associated with pesticides are insignificant compared to benefits of saving lives from these deadly diseases. The use of pesticides has increased our food supply and decreased cost of producing food, in addition to increasing the profit for farmers.

Pesticide Problems

Pesticides often kill nontarget species. Many times ecological pest controls are killed, like spiders, ladybugs, or wasps, which may result in an increase in pests because their natural predators are eliminated. The pesticide frequently kills honeybees and other insect pollinators, resulting in decreased fruit production. Some pests may survive a pesticide application due to a resistance to the pesticide, which will result in genetically resistant insects that can survive a subsequent application of the pesticide. This phenomenon is called pest resurgence. Farmers must administer pesticides in larger doses than before, known as the pesticide treadmill. The farmer tries to increase the insecticide dose but the resistance of pests requires increased doses of the pesticide, which requires more pesticides in a positive feedback loop. Occasionally a secondary pest outbreak occurs, such as when a pesticide kills off boll weevils but then secondary pests, such as the cotton boll worm, increase quickly due to the absence of natural predators. Pesticides tend to migrate through erosion, wind, and water. The grasshopper effect is seen when persistent chemicals migrate from the site used in warmer climates to condense and precipitate in colder regions, where bioaccumulation occurs in top carnivores such as polar bears, beluga whales, and humans. There are farm worker health problems associated with application of pesticides and working in fields where pesticides have recently been applied.

Relevant Pesticide Legislation

Three federal agencies regulate pesticides in food in the United States, the EPA, Food and Drug Administration (FDA), and the United States Department of Agriculture (USDA). The Federal Food, Drug, and Cosmetic Act (FFDCA) of 1938 allows the EPA to set tolerance levels for pesticide residues in food that are the concentrations in or on food that may pose an acceptable health risk. In 1958 the Delaney clause was added and states that no chemicals would be added in food if they cause cancer. The Federal Insecticide, Fungicide, and Rodenticide Act (FIFRA) of 1947 was enacted before the EPA existed, but now they regulate the sale and use of pesticides. In 1972 FIFRA was broadened to include registration of pesticides with the EPA. The EPA allows licensing of pesticides if they determine they will not pose significant risk to human health or the environment. The FDA and the USDA enforce pesticide use and tolerance levels set by the EPA. They can destroy any food shipments found to exceed EPA limits. In 1996, Congress passed the Food Quality Protection Act. The EPA can set pesticide tolerance levels taking into account the effects on aggregate pesticide exposures. The EPA has banned some chemicals previously used on fruits due the adverse effects on children. The chemical Dursban was also banned, based on an unacceptable risk to children's health. CCA, or chromated copper arsenate, treated lumber was banned for residential use in the United States due to the inherent danger of the arsenic in the wood.

In 2001, 127 countries agreed to ban persistent organic pollutants (POPs) including aldrin, chlordane, dieledrin, DDT, mirex, toxaphene, PCBs, dioxins, and furans. However, the production of these chemicals had continued in the United States. During that time, because the other countries had not banned the chemicals, they were sent back to the United States on imported crops, in a cycle known as the circle of poison.

Bhopal

In December of 1984, in Bhopal, India, the local Union Carbide plant, which produced Sevin, released many gases used to generate the pesticide, including methyl isocyanate. The plant was on the edge of a shantytown, so exact numbers of those affected are impossible to obtain. It is believed that 15,000 died immediately from the gas and over 800,000 people were exposed and suffered chronic disease, including lung disease, eye damage, immune dysfunction, nerve damage, cancer, birth defects, and impaired mental difficulty. Union Carbide paid India $470 million in compensation, but that money has not been distributed to the victims.

Organic Agriculture

For a food to be designated 100 percent organic, it must be produced without hormones, synthetic fertilizers, pesticides, or antibiotics. Nor may it be genetically modified. If a food is made with organic ingredients, it must contain 70 percent organic components. Organic foods tend to have more bacteria associated with them due to a lack of preservatives in the foods and an absence of antibiotics while rearing the livestock.

Integrated Pest Management

> Take Note: Many of the soil erosion control techniques are also valuable for pest control. You must be able to integrate your knowledge of soil techniques with those used in pest management to fully understand agricultural management and sustainability.

Integrated pest management (IPM) uses low doses of nonpersistent, nonbiomagnified pesticide to reduce pest numbers, followed by a combination of the following techniques as applicable for a specific pest. The goal of IPM is to decrease crop damage to an economic threshold where the potential damage justifies pest control expenditures. One of the easiest ways to control pests is by increasing crop diversity. Strip cropping or polyculture will make it more difficult for the insects to spread. Crop rotation keeps pest populations from building up, because the crop changes seasonally. Shelterbelts provide shelter for insect eating birds and other predators, as well as blocking easy movement by the insect. If an insect's life cycle is known, plantings can be coordinated around the expected date of hatching. Control of adjacent crops and weeds is essential, because nearby crops can harbor pests or they can use them as trap crops, which will attract insect pests from more valuable crops. For example, cherry peppers can be planted around a more valuable plot of bell peppers, or marigolds can be planted around a crop of garlic.

Natural predators and parasites, biological control, can be employed. For example, spiders eat a variety of insects, lacewings and ladybugs control aphids, and wasps lay eggs on caterpillars that hatch and then devour the caterpillar. Herbivorous insects may be used to control weeds. The biopesticide bacterium *Bacillus thuringiensis* (Bt) can be used to control leaf eating caterpillars. Bt has a toxic crystal that paralyzes the caterpillar digestive system and causes it to rupture. Sterile male insects can be released to mate with female insects. The females usually only mate once and by mating with the sterile male, her eggs are not fertilized, resulting in no offspring. Sterile males are useful in controlling Mediterranean fruit flies, tsetse flies, and screwworms. Genetically resistant strains of crops and animals are being developed as artificial selection continues. Pheromones, chemical signals between insects, can be used as sex attractants to catch males in traps or can be sprayed during breeding season, known as the confusion technique. These pheromones are highly species specific but expensive to develop. Insect hormones may also be used. Juvenile hormone prevents an insect from maturing, and molting hormone can cause a premature or incomplete molt.

Chapter 10 Questions

Use the following choices for questions 1–4.
    a. sterile male technique
    b. Bt spray
    c. ladybugs
    d. crop rotation
    e. trap crops

1. pest control method most effective on a Mediterranean fruit fly outbreak on citrus trees in Florida
2. biological control method preferred for combating aphids
3. cultural method effective for a variety of grain and legume crops when it is desired to sell all crops produced
4. pest control measure most effective on butterfly and moth larvae

5. Which of the following is the best reason to use organophosphates instead of organochlorines as a pesticide?
a. less persistent in the environment than organochlorines
b. more toxic to humans than organochlorines in lower doses
c. cost less than organochlorines to produce
d. bioaccumulate and biomagnify in many ecosystems
e. adversely affect the endocrine system while organochlorines only damage the nervous system

6. Which of the following is not a drawback to using insecticides to control a pest outbreak?
a. Many other insect species are killed by the pesticide, thus destroying natural predators.
b. Natural pollinators may be killed thus preventing the crop from producing fruits.
c. The insecticide can potentially protect the crop from damage due to the insect pest and thus improve land productivity.
d. The organisms may survive the pesticide application due to resistance, which will cause the subsequent insect population to resist the pesticide.
e. Some birds may be killed during the application because they tend to be sensitive to some pesticides.

7. Which of the following is an organochlorine pesticide?
a. DDT   b. arsenic   c. malathion   d. Sevin   e. Roundup

8. Which federal legislation requires companies to register their pesticides with the EPA?
a. FIFRA   b. Delaney clause   c. FFDCA   d. Food Quality Protection Act

9. Insecticides cannot help prevent the transmission of which of the following diseases?
a. dengue   b. malaria   c. African sleeping sickness   d. yellow fever   e. tuberculosis

10. Which of the following applications requires the least amount of pesticide for maintenance?
    a. algaecides to a pond exposed to cow feedlot wastes
    b. herbicides to a no-tillage field
    c. pesticides to a golf course
    d. insecticide inside a home to control pests
    e. insecticides to a field of cotton

Explanations
1. a. Sterile males are preferred because Bt and ladybugs would not work well on a fruit fly infestation. Crop rotation and trap crops likely would not work because the affected plants are trees.
2. c. Ladybugs are a biological control method that would work well to destroy an aphid infestation.
3. d. Crop rotation and trap crops are the only cultural methods that were choices. Trap crops are not sold; they are sacrificed to pull the pests away from the money-making crop.
4. b. Bt spray works well for caterpillars because the crystal that the bacterium makes kills the larvae.
5. a. Organophosphates are far less persistent in the environment than organochlorines, thus making them a preferred pesticide because they do not bioaccumulate or biomagnify. They are more toxic than organochlorines in lower doses, but that is a drawback to using them. They cost more than organochlorines to produce. Both organophosphates and organochlorines can adversely affect the nervous system.
6. c. The insecticide can protect the crop and improve land productivity. A farmer will use the pesticide, as the benefits outweigh the costs.
7. a. DDt is an organochlorine pesticide. Arsenic is an inorganic pesticide. Malathion and Roundup are organophosphate pesticides, and Sevin is a carbamate pesticide.
8. a. FIFRA requires that companies register pesticides with the EPA.
9. e. Tuberculosis is not a vectorborne disease; therefore insecticide cannot lower its rate of transmission.
10. d. All choices except applying insecticide inside a home to control pests would require an intensive amount of pesticides.

Chapter 11 – Biodiversity

Key Terms

biodiversity   intrinsic value
biodiversity hot spots   invasive species
CITES   overharvesting
endangered species   threatened species
extinction   umbrella species
flagship species   vulnerable species
habitat conservation plans

Skills
1. Relate the importance of biodiversity to humans.
2. Examine the negative impacts that humans have on biodiversity.
3. Summarize the legislation/policies (national and international) that protect species from extinction.
4. Explain the different mechanisms by which humans are assisting various species in recovering their population size. Include specific examples of species that have had population increases due to human intervention.

Biodiversity

Biodiversity is the variation of organisms in a given area. It includes species diversity, the number of different species in an area; genetic diversity, the genetic variety within a species; and ecosystem diversity, which is the complexity of species interactions in biological communites. Species diversity includes species richness, which is the total number of species in an area, and species evenness, the relative distribution of the different species in an area.

Definition of a Species

The definition of the word species is a group of similar looking organisms that can interbreed and produce fertile offspring. For example, because horses and donkeys produce the sterile mule when mated, they are different species. A more evolutionary-based definition is the phylogenetic species concept, which examines the relationships between organisms and their ancestry to determine if organisms are separate species. For example, for years guinea pigs have been classified as rodents. Recent genetic studies suggest that guinea pigs likely diverged from the rats and mice far earlier in evolutionary time than was realized, thus making them less genetically similar to rodents than previously expected. The evolutionary species concept is defining species based upon their evolution and their history.

Scientists are more able to determine relationships between organisms with the advent of molecular biology and the ability to analyze DNA to test relationships between organisms thought to be separate species. An example of a confusing species situation is the red wolf.

Many scientists are not certain if the red wolf is an endangered species due to its low population or if the reason they have a low population is because they are a hybrid species resulting from a breeding between a gray wolf and a coyote. Mitochondrial DNA (mtDNA) analysis suggests that the red wolf is a hybrid between the two species, because the mtDNA from the red wolves can be found in both the gray wolf and the coyotes. Nuclear DNA studies show no unique alleles in the red wolf, but alleles found in both of the other species. The data thus suggests that the species is a hybrid, but the time of the hybridization is not well understood, therefore classifying the species as separate seems prudent at this time. The species could have hybridized long ago, thus forming a new species. The (USFWS) has classified the red wolf as endangered, and the species has federal protection. The taxonomy is still confusing, because no definitive answers have been elucidated.

Number of Species on Earth

Figure 11.1 The chart estimates that there are approximately 1.5 million species known to be on earth. Taxonomists estimate that the actual number of species on earth is anywhere from 3 to 50 million total species.

Bacteria and cyanobacteria	4,000
Protozoa (single-celled animals)	31,000
Algae (single-celled plants)	40,000
Fungi (molds, mushrooms)	72,000
Multicellular plants	270,000
Sponges	5,000
Jellyfish, corals, anemones	10,000
Flatworms (tapeworms, flukes)	12,000
Roundworms (nematodes, earthworms)	25,000
Clams, snails, slugs, squids, octopuses	70,000
Insects	1,025,000
Mites, ticks, spiders, crabs, shrimp, centipedes, other noninsect arthropods	110,000
Starfish, sea urchins	6,000
Fish and sharks	27,000
Amphibians	4,000
Reptiles	7,150
Birds	9,700
Mammals	4,650
Total	1,733,000

Source: *Norman Myers, 2000.*

Biodiversity Hot Spots

The largest proportion of species on earth tends to be concentrated around tropical regions. Biodiversity hot spots are areas that have 1,500 endemic (native) species and have lost 70 percent of their habitat due to habitat disruption and invasion of exotic species. These 34 hot spots have 1.4 percent of the world's area but 75 percent of the world's threatened mammals, birds, and amphibians. They tend to be tropical islands, such as Madagascar, Indonesia, and the Philippines. The geographic isolation of these areas has resulted in highly unique flora and fauna.

> **Take Note:** You must be able to present an argument for maintaining biodiversity. Conversely, you may be asked to explain why a species should be allowed to become extinct. Be sure that you can present valid scientific arguments for both sides of the question. Do not interject your opinion, but relate a series of cogent statements to support the position you are asked to defend.

Utilitarian Benefits of Biodiversity

Primitive societies use wild plants and animals to provide clothing, food, and shelter. Nearly all of today's food crops have been cultivated through years of selective breeding of wild plants. Numerous plant species are used to derive medicines. For example, the rosy periwinkle provides a leukemia drug, and Pacific yews are used to produce the drug taxol to treat breast cancer. The cinchona tree produces quinine to treat malaria, and the foxglove provides digitalis, a heart drug. Plants and animals also have industrial uses, including clothing, waxes, fragrances, and dyes. Plants are also necessary for lumber, paper, and firewood.

Ecological Benefits of Biodiversity

The earth's biodiversity plays several necessary ecological roles. Plants function in the carbon cycle by carrying out photosynthesis, which produces the carbohydrate biomass necessary to feed the earth's herbivores. The other biogeochemical cycles are dependent upon the biotic portion of the cycle to maintain their function. Animals are exceptionally important in pollination, which allows fruit formation to support numerous animal species. Soil formation, pest control, waste disposal, and water purification are all dependent upon the species that make up the earth.

Nonutilitarian Benefits of Biodiversity

Nature has aesthetic value, meaning that the plants and animals on earth are beautiful. Humans enjoy interactions with nature, including hunting, fishing, bird watching, and scuba diving. Abundant biodiversity improves our quality of life. Ecotourism, tourism based upon examining nature, can generate tremendous income in various unique habitats throughout the world. The world's plants and animals also possess intrinsic, or existence, value because they have value simply because they exist.

Extinction

Extinction is the elimination of an entire species when the last of its members have died. It is an irreversible process. It is also a perfectly natural process, as species become less suited over time to changes in the environment. Species arise through mutation and natural selection, so those not as "fit" become extinct through the same processes. The background extinction rate is the natural rate of extinction. Mass extinctions have occurred periodically during earth's history, which wiped out vast numbers of species in a relatively short time by geological standards. After the mass extinction, radiation occurs where species rapidly evolve to fill empty niches in ecosystems. A well-studied example is the mass extinction of the dinosaurs at the end of the Mesozoic era, which led to radiation of mammals in the current Cenozoic era. Current theory indicates that these mass extinctions are due to climate change. Humans are currently accelerating extinctions by damaging habitats. Scientists consider the earth to be in a period of mass extinction at this time, because the current extinction rates are from 100–1,000 times background extinction levels. It is the first time in history of earth that one species is causing the extinction of another species. The primary threats to biodiversity are habitat disruption, invasive species, pollution, population, and overharvesting.

Historic Period	Time (Before Present)	Percent of Species Extinct	Figure 11.2 Mass Extinctions
Ordovician	444 million	85	
Devonian	370 million	83	
Permian	250 million	95	
Triassic	210 million	80	
Cretaceous	65 million	76	
Quaternary	Present	33–66	

Source: W. W. Gibbs, 2001.

Take Note: If you are asked to explain on an essay how species are impacted by humans, habitat disruption is never an answer. You must explain how the habitat is disrupted, the purpose of the disruption, and the impact the disruption has upon the animal species. For example, saying "The mountain gorilla is endangered due to habitat disruption" is not enough information. Stating that "The mountain gorilla is losing important food resources due to humans clearing the land for farming and grazing livestock" would state the impact on the animal, as well has why the land was being disrupted. A clarification of your statement will garner the points that are available on the essay.

Habitat Disruption

The two main reasons that habitat is disrupted is for agriculture and timber cutting. Land is cleared not only to grow crops but to allow rangeland formation for livestock. Timber harvest provides firewood, charcoal, lumber, and pulp. Habitats may also be disrupted as a result of construction and commercial development. Humans drain wetlands and flood other areas with water diversion projects. Mining disrupts significant amounts of habitat, particularly surface mining.

Exotic Species

Exotic species are considered biotic pollution because they are the introduction of a non-native species into an area where it is not natural. The species may be accidentally (zebra mussels via ballast water in the Great Lakes) or deliberately (melaleuca and Australian pine in South Florida to protect canals, levees, and other areas from erosion) introduced. These species are a problem when they become an invasive species due to a lack of controls on population growth in the new area. The areas most susceptible are tropical areas that have high levels of tourism and ports. Islands are extremely susceptible to damage by exotics because most islands have unique flora and fauna that may not be able to outcompete the invaders. For example, many of the Galapagos Islands have introduced cattle, goats, and sheep, large herbivores not native to the islands. There are several solutions to removing invasive exotics. All solutions can prove to be expensive and time consuming in the areas faced with having to eradicate these species. Mechanical controls, in which the exotics are physically removed, are an option. For example, Brazilian pepper trees or kudzu may be physically removed from a site to allow native species to flourish. Biological controls, where a predator or parasite is introduced to control the pest, may also be used. For example, the pink hibiscus mealybug, a pest that feeds on numerous types of fruit trees, is parasitized by several wasp species and eaten by ladybugs. Ecological controls, such as fire, may also eradicate exotics. Fires may kill exotic grasses and allow fire resistant native species to proliferate.

Pollution

Humans pollute air, water, and soil on a regular basis. Pesticides, PCBs, dioxins, mercury, and lead are just a few of the pollutants introduced by humans into the environment. Water pollutants include sediment, organic wastes, and nitrates and phosphates. Additionally, the regional problem of acid deposition and the global problems of ozone depletion and global warming are having adverse effects on a wide variety of ecosystems throughout the world. The chart at the end of the chapter reviews gives a fairly detailed listing of many of the pollutants important in the study of environmental science.

Human Population

The ever-increasing human population affects biodiversity as humans are forced to move into previously uninhabited areas. To support the huge population, we must harvest more timber, raise more grain and livestock, harvest more fish, and mine for more nonrenewable resources than ever before in the history of humans. All of these activities disrupt habitats, and thus have a profound impact on natural populations.

Overharvesting

Overharvesting has led to the endangerment and extinction of numerous species in the world. The American passenger pigeon became extinct in 1914 because they were commercially hunted and their food supply dwindled as they lost habitat due to deforestation. American bison were killed in the American west in a genocide attempt to destroy Native American populations dependent upon the bison for their livelihood. Several fish species are under severe pressure from overfishing, including sharks, bluefin tuna, and marlin. Animals may also be harvested in the wild for the pet trade, zoos, research labs, and aquaria. Numerous species of bird, including the yellow-crested cockatoo and the hyacinth macaw, are endangered or threatened due to the wild bird trade.

Take Note: Students have been asked on two previous essays about the characteristics of endangered species. They have also been asked to name an endangered species and explain the factors behind its endangerment. Be certain that you are familiar with several endangered species and the facts leading to their endangerment. Additionally, you must be familiar with legislation and policies that protect endangered and threatened species.

Endangered and Threatened Species

An endangered species, as defined by the Endangered Species Act, is a species in imminent danger of extinction throughout all or most its range. Examples of critically endangered species include the California condor, whooping crane, giant panda, and black rhinoceros. A threatened species is one whose population is likely to become endangered in the near future, throughout all or a significant portion of its range. Threatened species include the bald eagle, grizzly bear, and the American alligator. Vulnerable species, rare species, or species of special concern are potentially threatened species. These species usually have a measure of protection from harvest, which might include bag limits, size ranges, and seasons for harvesting these species from nature.

Characteristics of Endangered Species

Typically, endangered species occupy a narrow habitat or occupy a small range. Therefore, they are far more likely to become extinct if their habitat is altered. An example would be the various species of endangered prosimians, such as the aye-aye or ring-tailed lemur endemic to the island of Madagascar. This narrow habitat also reiterates that species that are native to certain islands are more likely to be threatened, such as the giant Galapagos tortoise, Galapagos hawk, and the Galapagos penguin. Animals that require a large territory, such as a predator, are likely to become extinct if their territory is modified by humans. Animals with low reproductive success, low reproductive rates, or small numbers of offspring per gestation are more likely to be endangered, such as the giant panda or African elephant. Animals that breed in very specialized areas, such as the whooping crane, may also suffer greatly from habitat disruption. Animals that have highly specialized feeding habitats, such as the giant panda's requirement for bamboo, make them subject to endangerment. Animals with a large size may also be endangered, as they require more space to supply them with their food needs. Examples include grizzly bears, gray wolves, and bald eagles.

Other aspects of endangered species are not a characteristic of the species themselves, but a characteristic in which humans find utilitarian or of economic value. Some species are destroyed by humans for meat; use in medicines; or as a prize, such as tigers, whooping cranes, sea turtles, whales, mountain gorillas, and rhinoceroses. Some species have other economic value, such as the ivory from an African elephant, the fur from ocelots or snow leopards, or the feathers from kakapos or roseate spoonbills. Many of the large predators have been eradicated by humans out of fear for themselves or concern for livestock.

Legislation

By the end of the 1800s, many states had enacted hunting and fishing restrictions to protect game species. The first federal legislation was the Lacey Act of 1900 which required a federal permit to transport live, dead, or parts of wild animals across state lines. This act was to help species protected in one part of the country but that may not be protected in other parts. For example in Florida and Mississippi the black bear is threatened and may not be hunted, but in Alabama and Georgia they are considered a game species.

In 1973, Congress passed the Endangered Species Act (ESA). The act protects endangered species from harm, harassment, hunting, shooting, killing, capture, or collection. It is illegal to import or trade in any product made of endangered or threatened species. There are 1,300 species on the endangered/threatened species list in the United States. The act authorizes the U.S. Fish and Wildlife Service (USFWS) to protect from extinction any endangered and threatened species and the National Marine Fisheries Service (NMFS) to do the same with marine species. The act requires USFWS and NMFS to prepare recovery plans and protect critical habitats for the species. The United States spends about $150 million per year in these recovery efforts. Nearly half of the money is spent on "charismatic" species, or ones that attract the attention of most people, such as the California condor or Florida panther. The lowly invertebrates and plants get less than $5 million per year toward their recovery. Funding priorities are frequently based on emotion, not biological impact of endangered species. Keystone species (bats or alligators) and indicator species (brook trout) merit special attention, because their role in the environment is more important than may be evident to the casual observer. Umbrella species, such as the tiger or grizzly bear, need vast amounts of undisturbed habitat to maintain Minimum Viable Population (MVPs). Flagship species are interesting or attractive so people respond emotionally to them. Examples of flagship species include the West Indian manatee and the giant panda. Successful recovery plans include those for the American alligator, bald eagle, and brown pelicans. Opponents of the act claim that it impedes economic progress because it tends to limit development. They also vehemently oppose that the act is only supposed to consider biology when placing species on the list, not the economic ramifications of listing a species. When an endangered species is on private land, the species is still protected. Therefore the presence of an endangered species could prevent the development of the land. As a result, the USFWS has helped devise habitat conservation plans, which will allow development of part of the land as long as the species is benefited overall. The ESA expired in 1992. Since that time, Congress has been debating whether to renew, weaken, or strengthen the act.

In 1996 Congress passed the Sustainable Fisheries Act (SFA). The SFA amended the Magnuson Fishery Conservation and Management Act of 1976 and it became the Magnuson-Stevens Fishery Conservation and Management Act. The Magnuson Act allowed the federal government to manage fisheries from 3 (9 for the Gulf coast of Florida and Texas) to 200 nautical miles off U.S. coast. They can impose limits on fish taken in this area. The revisions to the act called for the NMFS to implement management and conservation plans. Fish managed under the act include Atlantic sharks and swordfish, billfish, and bluefin tuna.

In 1975, the international agreement known as the Convention on International Trade in Endangered Species of Wild Flora and Fauna (CITES) was established. Its goal is to ensure that trade of species does not lead to their extinction. It bans hunting, capturing, and selling of endangered or threatened species. Species are listed in three appendices, with Appendix I species being endangered and in which trade is prohibited and Appendix II species are threatened and limited trade is conducted. Appendix III species are protected in one country and CITES assists in regulating the trade of those species.

Species Survival Plans

Many animals are taken from nature into captivity in zoos and bred in the attempt to increase their numbers, known as ex situ conservation. It is essential to determine what brought these species to the brink of extinction prior to trying to reintroduce the species to the wild. Fairly successful captive breeding programs have been established for the black-footed ferret and the whooping crane in the United States. Techniques used in captive breeding to increase the genetic diversity include artificial insemination, cross incubation, embryo transfer, and artificial incubation. Artificial insemination is the transfer of sperm without intercourse. One male can provide sperm to several zoos without having to be transported. Cross-incubation is allowing a bird of one species to hatch and care for a similar species. Unfortunately, this situation frequently backfires due to imprinting by the fostered animals. Embryo transfer is the production of numerous amounts of eggs in a viable female with the use of hormones. These eggs are harvested and fertilized in vitro. The embryos are then transplanted into less desirable genetic stock to raise the offspring. Artificial incubation is harvesting eggs in nature and rearing the offspring to a viable size in captivity to return to nature, such as Kemp's Ridley sea turtles. Botanical gardens serve the same purpose for assisting with the recovery of plant species. It is equally important to set aside habitat for critically endangered species in the wild to maintain as much genetic diversity as possible, known as in situ conservation. Protection from poaching is essential in many of these habitats.

Chapter 11 Questions

Use the following policies or acts for questions 1–4.
  a. CITES
  b. Endangered Species Act
  c. Magnuson-Stevens Fishery Conservation and Management Act
  d. Lacey Act
  e. Wilderness Act

1. prevents long-lining in U.S. coastal waters
2. the U.S. Fish and Wildlife Service created a management plan to protect the gray bat
3. international agreement that bans the trade of ivory from African elephants
4. two men from Idaho were convicted of violating this act after removing a bighorn sheep skull and horns from Yellowstone National Park

5. All of the following are characteristics of endangered species except
a. small ranges.   b. large body size.   c. few offspring per breeding season.
d. a highly specialized diet.   e. specialized feeding habitat.

6. The following best describes people's interest in endangered green sea turtles?
   I. They are a charismatic species because sea turtles attract the attention of most people since they are unusual creatures.
   II. They are a flagship species because people respond emotionally to them.
   III. They are an indicator species because they are endangered.
a I only   b. II only   c. III only   d. I and II   e. II and III

7. The bald eagle has had population resurgence since DDT has been banned, and the species is protected from harm. The eagle has been downgraded from endangered to
a. species of special concern.   b. threatened.   c. extinct.
d. rare species.   e. vulnerable species.

Questions 8–10 refer to the following paragraph and graph.

Whooping cranes migrate from Wood Buffalo National Park in Canada to Arkansas National Wildlife Refuge in Texas. The species is critically endangered as evidenced by the graph below.

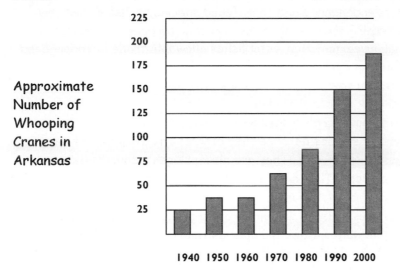

Approximate Number of Whooping Cranes in Arkansas

1940 1950 1960 1970 1980 1990 2000

Year

8. The graph above indicates that the whooping cranes
   a. had a population recovery without the help of humans by 1950.
   b. had numbers so low that they would have become extinct within the next century without the help of humans.
   c. have continued to decline in number regardless of human intervention.
   d. have become extinct in the last decade.
   e. nest in the Arkansas Wildlife refuge only every 25 years.

9. All of the following techniques are used to improve wild whooping crane numbers except
   a. captive breeding.   b. cross incubation.   c. artificial insemination.
   d. embryo transfer.   e. artificial incubation.

10. All of the following explain why whooping cranes are endangered except they
    a. fly into power lines as they migrate.
    b. were overhunted for feathers to adorn hats.
    c. can change the destination of their migration route on a yearly basis to avoid anthropogenic hazards.
    d. lost critical nesting habitat due to drainage of wetlands for agriculture.
    e. they produce two eggs per mating season, but only one egg is viable.

Explanations

1. c. The Magnusun-Stevens Act sets fishing standards from 3-200 miles off of the U.S. coast.
2   b. The ESA not only protects endangered and threatened species, it also makes the USFWS responsible for increasing their numbers by creating management plans.
3. a. CITES is the international agreement that establishes allowable trade in endangered and threatened flora and fauna.
4. d. The Lacey Act was violated on several issues because the men took a protected species skull from a national park across state lines.
5. a. Endangered species tend to have large ranges or territories.
6. d. Sea turtles are a charismatic species because they attract the attention of most people because they are unusual creatures. They are also a flagship species because people respond emotionally to them. They are not an indicator species, because they do not seem to be indicating environmental change.
7. b. Threatened species are those at risk of becoming endangered. Species of special concern, rare species, and vulnerable species are species that may become threatened in the near future. Extinction is the disappearance of a species.
8. b. The whooping cranes would likely have become extinct because their numbers were so low. Their population was not recovering and they did improve with human intervention. The birds have improved in number, and there is no evidence that they only nest in Arkansas every 25 years.
9. d. Embryo transfer is used only on species that support the offspring in the body of the female. Since whooping cranes lay eggs, they could not use embryo transfer.
10. c. Migration routes are not able to be altered as they are part of a bird's innate behaviors. All of other reasons are valid reasons why whooping cranes are endangered.

Chapter 12 – Land Use: Forests and Grasslands

Key Terms

clear cutting
closed canopy
continuous grazing
debt for nature swap
desertification
extractive reserves
forest management
fuelwood
industrial timber
land reform

mixed perennial polyculture
monoculture forestry
old growth forest
open canopy
pasture
rangeland
rotational grazing
selective cutting
swidden agriculture

Skills
1. Characterize the different types of forests and the products derived from them.
2. Summarize forest harvest techniques. Evaluate which techniques are better for the restoration of the habitat.
3. Describe the use of rangeland and pasture for raising livestock.
4. Relate the overuse of rangeland to the formation of deserts.

Land Use

The diagram below illustrates the use of land in the world. The portion that is 33 percent of the land use includes cities, arid areas that will not support agriculture or grazing, and wetlands.

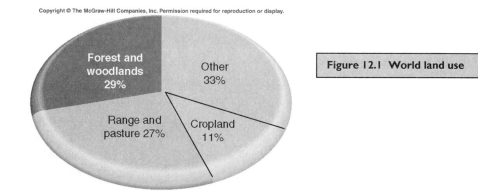

Figure 12.1 World land use

Forests

Closed canopy forests are those that have tree crowns that cover 20 percent or more of the ground. Open canopy forests have tree crowns that cover less than 20 percent of the ground. The forests of greatest concern to biologists are the old growth forests. An old growth forest is a forest in which there is little to no human intervention, and a tree can live out its entire lifespan. In a coastal redwood forest, that time could be thousands of years, and trees could reach 300 ft tall. In a temperate deciduous forest or southern pine forest, the time could be 300–400 years.

The percent of remaining old growth forests in different locales is widely varied. The largest remaining old growth forests are the boreal forests in Russia and Canada and the tropical forests in Brazil and Oceania. Secondary forests are those that have been cut and allowed to regenerate. Much of the eastern U.S. forest was cut when Europeans arrived in North America. Those forests have had 200 years to regenerate via secondary succession, and thus in many areas resemble old growth forest.

Utilitarian Forest Use

Humans use forests to harvest lumber for industrial timber, which forms lumber, plywood, veneer, and particleboard. Younger trees are harvested for wood pulp to make paper. Most of the industrial logging in the United States occurs in forests carefully managed for sustainable yield, not in old growth forests. In developing countries forests are frequently harvested for fuel wood (about 50 percent of all wood harvested worldwide) and to make charcoal. Developing countries frequently sell their exotic tropical hardwoods such as mahogany and teak to developed countries. These trees are frequently harvested from virgin old growth forests at an unsustainable rate. Forests are also damaged due to mining for nonrenewable resources and are clear-cut for rangeland.

Ecological Role of Forests

Forests serve as habitat for a huge number of species, thus additionally generating utilitarian recreational benefits for humans, including hunting, bird watching, and camping. Forests influence microclimate because they are typically cooler than the surrounding landscape. The air is cooled by evapotranspiration from the vegetation and the trees allow for the water cycle to function properly. Forests play a critical role because they serve as a watershed. Forests are also an integral part of the other biogeochemical cycles, particularly the nitrogen and carbon cycles. They absorb tremendous amounts of carbon dioxide, and thus function as a carbon sink. The roots in the forest function to decrease erosion and sediment runoff.

Forest Management

Forest management is the science of maintaining a forest for sustainable harvests, and is particularly involved in forest regeneration. Even-aged management is when all of the trees are the same size and age. This is important for conifers, as they require abundant light to grow adequately. If the trees are of a single species, monoculture forestry is occurring. The drawbacks to monoculture forestry include increased pests and disease and more erosion due to clear-cutting by timber companies. Monocultures degrade habitat as different species are adapted to a variety of trees in a forest. Uneven aged management permits natural regeneration by leaving some trees in the area. The trees are a variety of sizes and ages, which allows for more animal diversity, fewer diseases and pests, and less erosion since clear-cutting is not occurring.

Tropical Forests

Tropical forests are vanishing at an alarming rate. These forests include not only the tropical rain forests, but the tropical dry forests. The UN Food and Agriculture Organization estimates that 0.8 percent of the remaining tropical forests are cleared each year. The highest rates of deforestation in the world are in the Amazon and Congo River Basins. The major reasons for deforestation include agriculture, commercial logging, charcoal formation, grazing, oil and gas exploration, mining and smelting, and hydroelectric dams.

Indigenous people carry out subsistence agriculture in these forests. This type of agriculture is commonly called "slash and burn" agriculture, because the trees are felled and then burned to return nutrients to the soil. The correct term for such subsistence agriculture is swidden agriculture, because the farmer moves from plot to plot every few years, allowing the previous plots to lie fallow. The farmers plant fast growing crops to retain soil and use mixed polyculture to generate food and decrease pests. This method works sustainably on a small scale, but population growth and displacement of farmers from other areas have forced people to reuse the same plots repeatedly and the forest cannot regenerate.

Debt for Nature Swap

Debt for nature swaps allow developing countries to relieve their debt to developed countries and banks by setting aside protected areas of land in their country. By setting aside critical habitat, these nations are protecting the biodiversity in their country.

Temperate Forests

The two major issues in forest management in temperate forests are cutting old growth forests and methods of tree harvest. The temperate rain forests in the Pacific Northwest contain more biomass than any other biome. The biodiversity in these areas is tremendous, and numerous species, including the threatened marbled murrelet and northern spotted owl, require old growth forest as their habitat. Less than 10 percent of original old growth forest still exists in this area, and 80 percent of what is left will be cut down in the near future. Logging provides jobs to people in this area, and they are not willing to give up their livelihood. The compromise developed in 2000 was to leave 1 million acres of old growth forest and protect riparian areas necessary for salmon and other species dependent upon the rivers. Unfortunately the Bush administration revised the plan in 2003, eliminating any wildlife surveys and reducing the environmental reviews on logging.

Tree Harvest

All methods of tree harvest require that roads be built to transport the logs to the site of use. Roads result in increased erosion, increased sediment and debris in surface water, increased soil compaction, and habitat destruction and fragmentation. Streams and rivers are frequently damaged by silt and sediment that arises from tree removal. This damage is particularly profound in areas where salmon and trout spawn, because these species require clear water with rocky bottoms to lay their eggs.

The method of tree removal preferred by timber companies is known as clear-cutting. This method is used for most of the timber and pulpwood in the United States and Canada. Clear-cutting removes all of the trees from an area, regardless of size. This method is extremely detrimental to the environment because it eliminates habitat, induces erosion, causes the loss of mycorrhizae, and wastes small trees. Additionally, the use of the forest for recreation is prevented when the area is clearcut. Conifers flourish, because they require adequate sunlight to grow. Recall that these are early successional species, and thus this type of tree harvest works well. Clear-cut areas may undergo natural regeneration or may be replanted by timber companies.

Coppicing is leaving the stumps of a tree to regenerate the forest. Only some tree species can regenerate from a stump, including oak, maple, and ash. Seed tree cutting is cutting nearly all of the trees in an area, but leaving a few mature trees behind to provide seeds. The preferred method of harvest to an environmentalist is selective cutting, cutting only a small percentage of mature trees every 10–20 years while the rest of the forest remains intact. The area can then undergo natural regeneration via seeds from the remaining trees. This method allows for maintenance of biodiversity and causes less erosion than other methods. This is far more expensive for timber companies.

History of U.S. Forests

In 1897, Congress passed the Forest Management Act, which created the Forest Reserves to serve as a source for timber, mining, and grazing. The main goal was to provide cheap timber for home builders. In 1905, the newly renamed U.S. Forestry Service was moved from the Department of the Interior into the Department of Agriculture and the old forest reserves were relabeled as national forests. The Forest Management Act governed the harvest of the national forests until 1960, when Congress passed the Multiple Use Sustained Yield Act, which requires national forests to be managed for multiple uses with the result of a sustainable yield of products and services. Now the forests are not only to be managed for lumber production but must be managed for recreation and wildlife habitat as well. In fact, 75 percent of all forestry jobs involve recreation, and only 3 percent pertain to logging.

U.S. Forest Management

Most of the U.S. forests are managed by the USFS. The forestry service frequently sells logging rights for less than the cost of managing an area, which basically amounts to a subsidy for timber companies. Subsidized logging is expensive, primarily because of the building of roads through the areas to transport equipment and felled trees. The USFS builds and manages more roads than any other agency in the United States. The first biologist to head the USFS was Mike Dombeck, and in 2000 he initiated numerous changes that dramatically benefited U.S. forests. He prohibited roads in several areas, blocked sales of old growth forests, and included ecosystem management in forestry management programs. Most of his policies were reversed under President Bush's administration, as they opened more public land to logging, mining, oil and gas exploration and drilling, road building, and motorized recreation.

Fire

As discussed in chapter 5, fire is exceptionally important in a variety of ecosystems. It burns off the duff; returns nutrients to the soil, and prevents large canopy fires, if forests are maintained properly. Until 1972, it was the U.S. policy to quickly extinguish blazes in forests. The impact of allowing fire repression for so long allowed tremendous amounts of debris to build up, which resulted in catastrophic fires far more damaging to the forests than small ground fires. Therefore, the policy since 1972 is to let fires burn as long as areas will not be severely impacted. Foresters currently use prescribed surface fires to remove undergrowth on a regular basis to prevent crown fires. Bush's administration is using the "Healthy Forests Initiative" to legalize salvage logging. Any weakened or dead trees can be harvested in national forests, which will permit logging in previously unlogged areas. The last time salvage logging was legalized, numerous healthy trees were deemed diseased and removed. This salvage also does not address the materials that induce the hot fires, the accumulated duff and debris on the forest floor. Standing dead trees, or snags, are exceptionally important to the wildlife in a forest, as they serve as habitat.

Sustainable Forestry

Consumers can, by their shopping choices, alter the damage to forests. They can purchase sustainably produced wood products. Consumer demand can even alter large chain store policies. For example, Home Depot only buys wood products from suppliers that engage in environmentally friendly logging, so the $5 billion in lumber they sell each year then is sustainably raised. Staples, the office supply company, puts 30 percent recycled paper in their recycled products, up from their original 10 percent. Forests also do not have to be logged to remove resources. Forests can be used sustainably for extractive reserves or products such as nuts, fruits, mushrooms, latex (rubber), and chicle (gum) that do not require the trees to be removed.

Grasslands

Grasslands naturally support migratory herds of grazers, such as bison, gazelles, and antelope. These areas are not only used for agriculture, but they are also used as rangeland, which provides food for domestic animals. Rangeland differs from pasture, managed grassland or enclosed areas that support domesticated animals. Native grasses predominate in these areas and their dense sod works to decrease erosion. Perhaps the greatest problem associated with grassland management is having too many animals on a plot of land for too long, which results in overgrazing. In this situation, the animals will eat so much of the plant that the roots will die. The removal of these grasses contributes to erosion and soil compaction. Rangelands are located in grasslands with prolonged dry seasons, so desertification may occur. Nomadic herders avoid this problem by constantly moving their animals to avoid overuse of the area.

The predominate livestock grazed on grasslands are the ruminants, organisms with a chambered stomach that can use the cellulose in plant cell walls due to symbiotic bacteria in their gut. Only 15 percent of U.S. livestock feed comes from native grasslands. Ninety percent of the total U.S. grain crop, including corn, alfalfa, and oats, is used for livestock feed.

U.S. Rangeland

Sixty percent of the U.S. rangeland is privately owned. Only a small proportion of the cattle and sheep in the United States are grazed on federal land. Most of the federal rangeland is managed by the Bureau of Land Management (BLM) in the Department of the Interior. The USFS manages the rest of U.S. rangeland. Much of the federal land is in poor shape due to overstocking and lack of enforcement of stocking regulations. The government allows livestock to graze on federal rangeland for a small fee (the 2005 fee was $1.79/head/month). These prices are so low that they are well below the operating costs and management funds of the BLM, again resulting in a subsidized program.

Rangeland Management

The purpose of good range management is to permit sustainable grazing and to allow for the improvement of the range forage. One of the most important aspects of management is controlling the stocking rate, or number of animals, on the land. There are two main types of grazing methods. Continuous grazing results in animals rather quickly removing nearly all the vegetation in an area. Rotational grazing is allowing the animals to feed intensively for a short time, then moving them to allow the area to regenerate. This method allows regeneration as the grazed areas have fertilizer added from the feces, and seeds are distributed as the animals feed. This type of grazing also helps control weeds and protects grassland habitat. It is also beneficial to raise several species together, because cattle and sheep feed on grasses and herbaceous plants, but goats prefer forbs. In these managed areas, it is essential to protect the riparian areas because the animals tend to collect around the water and damage the banks.

Land Reform

Land reform movements attempt to redistribute land ownership from the few people who own the land to families who need the land to survive and who typically already inhabit the area. Many times land reform only occurs after revolution or war, which results in alteration of land ownership. Another concern in many nations is the role of the indigenous people in the area. These people are routinely exterminated as others enter into their native lands. Nations vary from those who feel indigenous people have no rights to those where there are indigenous rights given to large tracts of land.

Chapter 12 Questions

Use the map to answer questions 1-5.

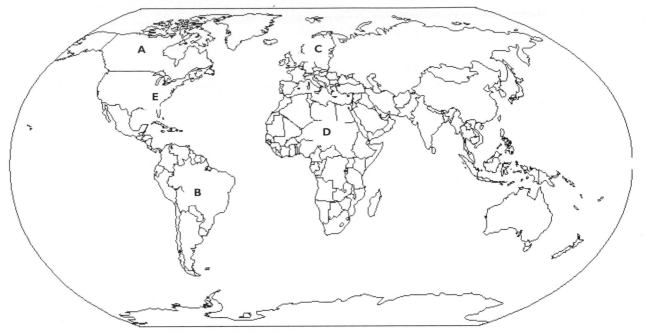

http://www.peacecorps.gov/wws/lessons/poster96a.html

1. large proportion of remaining old growth boreal forests
2. has lost the greatest amount of forested land compared to its original forests
3. less than 0.1 percent of the original forest still remaining
4. harvested for its tropical hardwoods resulting in tremendous loss of biodiversity
5. harvested by people needing firewood or charcoal

6. Problems associated with monoculture of trees include all of the following except
a. increased pests in monoculture pine forests.
b. decreased biodiversity on tree farms.
c. increased disease causing fungi in monoculture fruit trees.
d. a monoculture plantation could interfere with the water cycle in some locales.
e. increased sunlight helps promote growth of conifers in monoculture.

7. Which of the following best represents an ecological benefit of forests?
a. create a beautiful view for humans
b. can be used for hunting, fishing, and camping
c. provide wood pulp for paper
d. play an integral role in the water cycle
e. provide lumber for building homes

8. Which method of timber harvest is preferable for southern yellow pines, such as loblolly, longleaf, and slash pines, typically used for pulpwood to make paper?
   a. shelterwood cutting   b. clear-cutting   c. selective cutting
   d. coppicing   e. seed tree harvesting

9. Which of the following is a consequence of overgrazing?
   a. abundant riparian vegetation   b. increased sod and topsoil
   c. desertification of marginal lands   d. decreased erosion
   e. decreased soil compaction

10. All of the following can be harvested sustainably from forests except
    a. nuts.   b. minerals.   c. rubber.   d. fruits.   e. syrup.

11. All of the following are causes of deforestation of tropical forests except
    a. acid deposition.   b. logging.   c. mining.
    d. agriculture.   e. grazing.

12. For which of the following reasons would the USFS order a prescribed burn?
    a. remove snags   b. rid an area of pest species
    c. remove nutrients from the forest   d. clear the forest for agriculture
    e. help reduce erosion

13. Which of the following would be a good rangeland management practice in areas subject to a dry season?
         I. rotational grazing   II. deferred grazing   III. grazing multiple species
    a. I only   b. II only   c. III only   d. I and III   e. II and III

14. Which of the following best describes how U.S. forests are currently managed by the USFS?
    a. primarily for timber harvest
    b. primarily for timber harvest and grazing
    c. managed for timber, habitat, and recreation
    d. preserved for the aesthetic value to future generations
    e. preserved as species habitat and recreation sites

15. All of the following are reasons why federal rangelands are in poor condition in the United States except
    a. large numbers of bison present on federal land.
    b. overstocking animals on the rangeland.
    c. low subsidized grazing fees
    d. lack of enforcement of stocking rates.
    e. limited funds for range improvement

Explanations

1. a. Canada in North America has the greatest proportion of its original old growth forest and has the only boreal forest labeled in the diagram.
2. d. Africa has lost the greatest proportion of its original forest according to figure 12.10 in the textbook.
3. e. The temperate deciduous forests of North America have lost nearly all of their original area to agricultural land. Much of the forest is returning via secondary succession.
4. b. The South American rainforests have some of the greatest biodiversity in the world, yet the area is being harvested at an alarming rate.
5. d. Africa has a large population in need of firewood and charcoal to meet basic needs.
6. e. One of the reasons that conifers are grown using monoculture is because of the increased sunlight available to the trees after clearcutting. All of the other answers are definitely problems associated with monoculture of any type.
7. d. Forests playing a role in the water cycle is an ecological benefit of forests. All of the other choices are utilitarian or aesthetic value.
8. b. Clear-cutting because conifers grow best when they have full sunlight and removal of all trees in an area would accomplish this goal. Coppicing does not occur in pine trees. All of the other answers involve leaving some of the trees in an area.
9. c. Desertification can be a consequence of overgrazing. Typically riparian habitat is damaged, as are the sod and topsoil, which leads to erosion. Too many animals on one plot of land may also increase soil compaction.
10. b. Minerals are a nonrenewable resource and therefore cannot be harvested sustainably from forests.
11. a. Most tropical forests are not subject to acid deposition whereas they typically are logged for timber, have mineral and oil extraction, and are used for agriculture and ranching.
12. b. Prescribed burns are frequently used to remove pests, such as insects or fungi, from an infected area. The USFS would not burn to remove snags, as they serve as habitat. Burning would not remove nutrients from a forest, nor would a national forest be used for agriculture. Burning would more likely increase erosion than reduce it.
13. e. Rotational grazing, moving the animals around, would help reduce environmental degradation. The grazing of multiple species would help by reducing all plant species equally.
14. c. U.S. forests are managed by the multiple use sustainable yield principle, which manages the forests for timber, habitat, and recreation.
15. a. Although herbivores can increase competition and damage the range in general, there are not large numbers of bison present on federal land.

Chapter 13 – Preserving and Restoring Nature

Key Terms

corridors	re-creation	wilderness
ecosystem management	rehabilitation	wildlife refuges
floodplain	remediation	world conservation strategy
landscape ecology	restoration ecology	
reclamation	wetland mitigation	

Skills

1. Discuss the problems facing national parks.
2. Debate the value of national parks as recreational and aesthetically pleasing areas with their value in terms of economic development.
3. Explain the value of wildlife refuges, and explain the problems these locations are facing.
4. Characterize the principles of landscape ecology.
5. Justify restoration, mitigation, and replacement of natural areas damaged by humans.

U.S. National Parks

In 1872, President Grant established the first national park in the world, Yellowstone National Park. In 1912 the U.S. National Park System was established to protect areas with scenic beauty and cultural importance. In 1916 the National Park Service Act created the National Park Service in the Department of the Interior with Stephen Mather as the first director. The national park system (NPS) currently has 388 parks covering 280,000 km^2. These areas range from parks such as the Dry Tortugas National Park near Key West, Florida, historical sites such as Gettysburg National Military Park, battlefields like Antietam National Battlefield, memorials like the Korean War Veterans Memorial in Washington DC, and recreation areas such as Big Thicket National Preserve in Beaumont, Texas. Many of these parks are threatened with overcrowded conditions due to the large numbers of tourists that sojourn to them on a yearly basis. Tourists bring with them the additional problems of litter, increased traffic, and increased pollution. For example, personal watercraft (jet skis) and snowmobiles increase air and water pollution. Off-road vehicles induce compaction, erosion, and damage habitat. More federal funding is needed to combat these mounting problems. By placing tourists on buses to enter, several parks have avoided allowing cars. National parks have been carefully guarded in the past but nearby mine drainage and timber harvest have impacted these protected areas. Power plant emissions have created acid deposition that threatens many of these natural areas. The Bush administration opened 13 national monuments to oil and gas drilling.

Wilderness Areas

Areas of undeveloped land undisturbed by human activities and that humans visit but do not inhabit are considered wilderness areas. The Wilderness Act of 1964 allows public wilderness areas to be protected as part of the National Wilderness Preservation System managed by the NPS, USFWS, BLM, and USFS. The USFS continues to set aside areas that are roadless and uninhabited that may qualify for protection as wilderness. Reasons for maintaining wild areas include: quiet recreation areas, areas of scenic beauty, ecological research areas, refuges for endangered species, and the intrinsic value of the area in its pristine state. Most of the area currently under consideration as wilderness is managed by the BLM. As one might imagine, the timber companies, mining companies, and ranchers are more interested in the utilitarian value of these areas.

Wildlife Refuges

The National Refuge System was begun in 1901 by Theodore Roosevelt. There are currently 545 refuges in all 50 states, representing all major ecosystems in the United States. Wildlife refuges are managed by the USFWS, and these areas are managed for multiple use. Ironically these refuges are used for hunting, trapping, camping, snowmobiling, fishing, oil drilling, timber harvesting, mining, and grazing. These sites face pollution problems from these activities and external pollution is causing significant problems to much of the wildlife. The Arctic National Wildlife Refuge (ANWR) in Alaska is making the news because Congress is considering opening up the area to gas and oil exploration and drilling. The area is used as a caribou calving ground, and many environmentalists adamantly oppose using the area because the amount of oil that might be found is subject to debate. Alaskans are generally in favor of drilling in the site.

Global Parks

Many countries have been made aware of the increased threat to wildlife and natural systems. More than 100,000 protected areas now exist that protect 18.8 million km^2 of habitat. Much of this land has been protected using debt for nature swaps and more sustainable use of extractive industries of logging and mining. The International Union for the Conservation of Nature (IUCN) has developed the world conservation strategy to accomplish the following three objectives: maintain natural ecological processes, preserve genetic diversity in cultivated plants and domesticated animals, and sustainably use wild species and ecosystems. Some biomes have tremendous amounts of protection, including the boreal forests in North America and Europe. Some areas, such as the U.S. temperate grasslands and the U.S. chaparral, lack sufficient protection. Other areas in need of further protection to prevent escalation of ecological damage are the island of Madagascar, Papua New Guinea, and the Brazilian rainforest along the Atlantic Ocean.

Marine Systems

The United States protects its unusual marine ecosystems by creating sanctuaries to protect endangered species or shipwrecks of historical significance. The National Marine Sanctuary Program is administered by the National Oceanic and Atmospheric Administration (NOAA) as established by the Marine Protection, Research, and Sanctuaries Act passed in 1972. The program has 13 underwater parks, including the Florida Keys National Marine Sanctuary. Parts of the sanctuaries are deemed "no-take" zones, where absolutely nothing may be removed from the area—not even an empty seashell. These areas protect the species living in the area, but affect those species outside the area by providing food, shelter, and nursery space. Several international sites are in need of further protection, including the Philippines, Cape Verde Islands, and the Red Sea.

Conservation and Economic Development

Much of the land that needs to be protected throughout the world to alleviate potential endangerment and permanent damage to ecosystems is located in developing countries. These countries have difficulty managing their resources sustainably, because they lack the political and economic resources available to developed countries. The Man and Biosphere Program (MAB) encourages protection with sustainable use of outer areas. As shown in the diagram below, the core area allows full protection of the species located within it. The buffer zone allows humans to enter, but only for ecotourism and research. The outer areas are managed for multiple uses, including housing, roads, sustainable resource harvest, and agriculture.

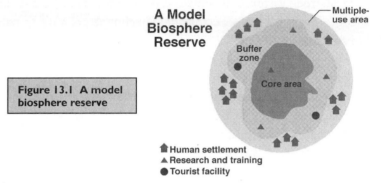

Figure 13.1 A model biosphere reserve

Landscape Ecology

Landscape ecology is the study of how the ecological patterns of an area shape the features of the land and in turn how the landscape shapes the ecosystem. Most biomes are made up of patches of different areas, which on close examination may give way to another area. These areas are like the bits of glass in a mosaic, because it takes all of the smaller pieces to see the larger biome. The major factors that shape landscape ecology are space and time.

Restoration Ecology

The size of a reserve must allow protection of endangered species, ecosystem sustainability, and protection from outside forces. If several small reserves are not contiguous, then corridors may need to be established to allow species to move from one area to another. These species can then exchange genetic material to maintain genetic diversity and prevent extinction due to geographic isolation. Restoration ecology is repairing ecosystems damaged by humans or natural forces. Restoration is bringing an area back to its former condition. Rehabilitation is helping rebuild an ecosystem, perhaps for human use, without restoring it to its original structure. It is not fully restored to its original state. Remediation is cleaning up contaminants or restoring a polluted area to its natural state. For example, sunflowers may be used to remove lead from the soil in areas contaminated with the heavy metal. Using plants to remove toxins is known as phytoremediation. Reclamation refers to restoring badly damaged sites, such as those subjected to mining, to its original topographic and ecological state. Re-creation is building a new community on a site so disturbed, restoration is not an option.

A good example of a restoration project is the Kissimmee River restoration. It is hoped this will reverse the damage done to the Florida Everglades. The Kissimmee River meandered through south

Florida, helping to create the famous River of Grass. The U.S. Army Corps of Engineers removed the meanders from the river to drain what they considered to be useless wetlands. The impact of the damage is far-reaching.

Restoration may involve intensive planting of native species and reintroduction of animals or may be as simple as letting the area naturally undergo succession and reach its natural state.

Wetland Preservation

Wetlands are extremely important components of the hydrologic cycle by allowing recharge of groundwater and purification of surface water. They also have tremendous biomass and may function as nurseries. The Clean Water Act of 1972 requires discharge permits to release wastes into surface water. In 1977, federal courts ruled that wastes also included filling of wetlands, but not drainage of wetlands. In 2001 the Supreme Court ruled that the act did not apply to isolated wetlands, ponds, or tidal mudflats. This ruling eliminated protection for 20 percent of U.S. wetlands. Wetland mitigation is creating new wetlands in lieu of those destroyed by development. A great example of wetland mitigation is Walt Disney World's mitigation banking in Central Florida. To be able to use certain portions of its property, Disney set aside over 8,000 acres as the Disney Wilderness Preserve to restore and protect wetlands and the 40 endangered and threatened species that inhabit those wetlands. A major problem with mitigation is that a simple pond might fill the legal requirement for replacement, but they are replacing natural swamps and meadows frequently damaged by human encroachment. Many farmers are paid conservation easements to allow wetlands to remain on their property.

Floodplains

Floodplains are the areas around a body of water that may flood during high water. These areas are low-lying areas adjacent to rivers and lakes and may be wetlands. These areas are rich in nutrients due to the periodic flooding that causes the nutrient-rich sediment to wash onto the floodplain. Humans have settled in these floodplains and established farms because the area is so nutrient rich. Unfortunately when flooding now occurs, human lives and property are at risk in these settled areas. To alleviate flooding, levees have been built to protect low-lying areas. In the summer of 2005, Hurricane Katrina damaged levees around New Orleans, Lousiana, so severely that vast portions of the city flooded, resulting in hundreds of deaths and billions of dollars in property damage and cleanup costs.

Ecosystem Management

Ecosystem management attempts to manage ecosystems with regard to ecology, economic, and social goals. The federal agencies that manage U.S. land have instituted ecosystem management practices in their individual management areas. These areas are to be managed for the entire ecosystem, not just one imperiled species. Natural units, such as watersheds, must be managed as a unit, not by jurisdiction. The ecosystem must be continually monitored to alter the management plan as needed. Humans are to be considered in these management plans, as they are a part of nature, not above it. Also, human values must play a role in this ecosystem management.

Chapter 13 Questions

Use the following choices for questions 1-5.

        a. mitigation
        b. remediation
        c. restoration
        d. reclamation
        e. rehabilitation

1. conversion of an agricultural field into a tall grass prairie by planting native grasses and reintroducing native herbivore species
2. returning a strip mine to its original topography and replanting with native plant species
3. using oil degrading microorganisms on an oil spill from a tanker
4. placing concrete structures near coral reefs to encourage growth of new reefs on the artificial structures
5. adding lime to a pond acidified by acid deposition

6. Which of the following is not a consideration in managing our national parks?
a. fire management   b. reintroduction of predators
c. ecotourism   d. rangeland management
e. education of visitors

7. Areas in greatest need of protection to preserve biodiversity include all of the following except
a. boreal forests.   b. marine ecosystems.   c. grasslands.
d. chaparral.   e. tropical rain forests.

Use the paragraph below for questions 8 and 9.

Interstate 75 in South Florida is known as Alligator Alley, because it runs through the Everglades, an area densely populated with a large variety of wildlife, the most noticeable being the American alligator. Due to the high number of animal deaths from automobile collisions, including the endangered Florida panther, the Florida Department of Transportation built fencing along the edges of the interstate. They also put in a series of underpasses under the interstate to allow animals to move from one side of the road to the other.

8. Florida's system would best be described as a series of
a. national preserves.   b. wildlife corridors.   c. buffer zones.
d. wilderness protection areas.   e. wildlife refuges.

9. The Florida panther is dependent upon these underpasses for all of the following except
a. maintaining genetic diversity to preserve the species.
b. preventing automobile collisions with the panthers.
c. establishing breeding habitat in areas traditionally inhabited by the Florida panther.
d. permitting access to an adequate food supply and adequate hunting territory for the large carnivore.
e. fenced areas prevent hunters from harvesting panthers because the animals may only be hunted in areas not bounded by the fences.

10. Wetlands are ecologically important because they
a. serve as habitat for waterfowl.　　b. have little species diversity.
c. cannot be used to grow crops.　　d. destabilize shoreline.
e. release water and result in flooding in low lying areas.

Explanations
1. c. Conversion of an agricultural field into a tall grass prairie by planting native grasses and reintroducing native herbivore species is restoration because the ecosystem is being returned to its natural state.
2. d. Returning a strip mine to its original topography and replanting with native plant species is reclamation because a badly damaged area is being repaired.
3. b. Using oil degrading microorganisms on an oil spill from a tanker is remediation because a pollutant is being removed from a natural ecosystem.
4. a. Placing concrete structures near coral reefs to encourage growth of new reefs on the artificial structures is mitigation because a new ecosystem is being established in an area where one had not previously existed.
5. b. Adding lime to a pond that has been acidified by acid deposition is repairing pollutant damage.
6. d. National parks are not used for rangeland, because they are protected areas in the United States.
7. a. Boreal forests have a tremendous amount of protection compared to marine ecosystems, grasslands, chaparral, or tropical rain forests.
8. b. Florida's system would best be described as a series of wildlife corridors because they allow the animals adequate space for survival and prevent them from being killed by auto collisions. None of the other terms describe the use of fences and underpasses to protect species.
9. e. As stated in the question, the Florida panther is endangered, and thus hunting is illegal. The animal is dependent upon these areas for maintaining genetic diversity, preventing automobile collisions, establishing breeding habitat, and permitting access to an adequate food supply.
10. a. Wetlands are ecologically important because they serve as habitat for waterfowl. Wetlands have high species diversity and can be used to grow crops such as rice and blueberries. Wetlands stabilize shorelines and protect from floodwaters by allowing water to slowly infiltrate into the groundwater.

Chapter 14 – Geology and Earth Resources

Key Terms

bauxite	magma	sedimentation
chemical weathering	mantle	smelting
converging plate boundary	mass wasting	subsurface mining
core	mechanical weathering	surface mining
crust	metal	tectonic plates
diverging plate boundary	metamorphic rock	transform boundary
earthquake	mineral	tsunami
heap leach extraction	mineral resources	volcano
igneous rock	rock	weathering
lava	rock cycle	
lithification	sedimentary rock	

Skills

1. Examine plate tectonics, distinguishing between transform, converging, and diverging plate boundaries and the geologic hazards associated with each.
2. Diagram the rock cycle, including the relationships between the three types of rocks.
3. Assess the importance of mineral mining and processing.
4. Compare and contrast the economic and environmental costs and benefits of mining and processing.
5. Classify the types of geological hazards.
6. Contrast the different types of weathering.

> **Take Note**: It is imperative that you are familiar with the earth's structure. You must be familiar with the plate tectonic theory and know what geologic hazards are associated with the different types of plate interactions.

Earth's Structure

The center of the earth, the core, is comprised primarily of iron with a small amount of nickel. The core is thought to generate the magnetic field associated with the earth. Surrounding the core is the mantle, the largest part of the earth's structure. The mantle contains large amounts of oxygen, silicon, and magnesium. The asthenosphere is the portion of the slightly molten upper mantle, which allows it to flow. The asthenosphere is the origin of magma. The solid upper mantle and lower portion of the crust form the lithosphere, broken into a series of plates, known as tectonic plates. The crust is comprised of continental and oceanic crust. Continental crust has large amounts of calcium, potassium, sodium, silicon, and aluminum. Oceanic crust composition is primarily igneous rock called basalt.

Plate tectonics

The tectonic plates are formed by the convection currents in the asthenosphere, which fracture the lithosphere into separate plates. There are three types of interactions at plate boundaries: divergent plate, convergent plate, and transform plate. At divergent plate boundaries, the plates slowly pull apart. Magma rises up to fill the void created when the plates separate. Therefore, the rock formed at these junctions is igneous rock. Divergent plate boundaries form mid-oceanic ridges in the oceanic crust. The mid-Atlantic ridge is an example of a divergent plate boundary. These oceanic mountains are far greater in size than any continental mountains. Convergent plate boundaries form when two plates collide. Typically a continental plate will rise above an oceanic plate in these areas, forming an area called a subduction zone, where the oceanic plate descends into the asthenosphere, forming more magma. Convergent plate boundaries therefore are usually indicated by volcanic activity. An example of mountains formed by a convergent plate boundary is the Andes Mountains in South America. Deep ocean trenches such as the Marianas Trench are also characteristic of subduction zones. The Ring of Fire and area of intense earthquakes and volcanoes in the Pacific Ocean are also associated with subduction zones. Transform plate boundaries occur when two plates slide sideways against each other. An example of a transform plate boundary is the San Andreas fault in California. Transform boundaries frequently experience earthquakes.

Rocks and Minerals

A mineral is a naturally occurring inorganic solid element or compound with a definite chemical composition and a regular crystalline structure. Examples of minerals include silicate minerals such as quartz or feldspar; carbonate minerals such as calcite or dolomite; sulfide minerals like galena or pyrite; sulfate minerals such as gypsum; and native element minerals that form gold, silver, diamonds, and graphite. Rocks are solid aggregates of one or more minerals. Rocks have characteristic composition. For example, granite is a mixture of feldspar, quartz, and mica crystals.

Rock Cycle

Weathering is a process by which rock is worn away to form individual particles. There are two major types of weathering. Mechanical weathering is the physical breakdown of rock without altering the chemical nature of the rock. Frost wedging is an example of mechanical weathering. Water expands when it freezes, so if rocks are penetrated by water that then freezes, the rock fractures into smaller pieces. Rocks abrading against other rocks in streams or in a glacier also cause mechanical weathering. Thermal expansion and contraction of rocks will also cause small pieces of the rock to break off. Organisms, including trees, worms, and rodents, can cause mechanical weathering by burrowing through the rock. Chemical weathering is the breakdown of a rock by altering its chemical nature. A good natural example of chemical weathering is the formation of soil by the growth of lichens on bare rock. Lichens secrete carbonic acid, which chemically degrades rock, forming soil. Other examples of chemical weathering include oxidation of rock (such as olivine weathering to form hematite) and hydrolysis (such as feldspar forming clay). Different rocks are susceptible to different types of weathering. For example, sandstone is susceptible to mechanical weathering, whereas limestone is susceptible to chemical weathering by acids.

Rocks are continually weathered to form particles that collect in a process known as sedimentation. These sediments may be compressed or compacted in a process called lithification to reform new rock. Rocks in subduction zones melt upon reaching the asthenosphere, only to re-emerge as igneous rock arising from magma. The process of rock formation, weathering, and lithification is known as the rock cycle.

Three major types of rocks arise from the rock cycle: igneous, sedimentary, and evaporite. Igneous rocks are the most common type of rocks on earth. They arise from cooled magma. Examples of igneous rock include basalt, granite, obsidian, and pumice. Sedimentary rock arises from the compression or compaction of sediments formed by weathering. Evaporite rocks form when a body of salt water dries up and the salt remains. Examples of sedimentary rock include sandstone, limestone, conglomerate, and the evaporate rock halite. Although technically formed by compression of ancient land plants, lignite and bituminous coal are considered to be biochemical sedimentary rocks. Metamorphic rocks are rocks formed from sedimentary or igneous rocks subjected to high heat and pressure. Examples of metamorphic rock include marble, slate, schist, and quartzite. Anthracite is a type of metamorphic rock, because it forms from lower grades of coal exposed to heat and pressure.

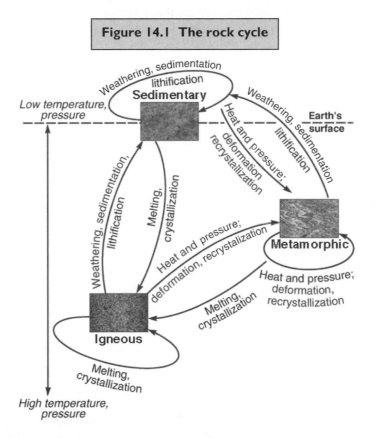

Figure 14.1 The rock cycle

Economic Geology and Mineralogy

Economic mineralogy is the study of minerals that have economic value to humans. Many valuable minerals are metal bearing ores, such as bauxite for aluminum or hematite for iron. The metals most used throughout the world are iron, aluminum, manganese, copper, chromium, and nickel. The greatest users of the metals are the United States, Japan, and Europe. Nonmetallic mineral resources include mica, asbestos, limestone, gravel, sand, graphite, quartz, diamonds, and other gemstones. Sand and gravel production combine to form the greatest proportion in volume and economic value of all of the nonmetallic mineral resources. Sand and gravel are used in bricks and concrete. Silica sand is the basis of glass. Limestone is quarried for concrete, road rock, and building stone and is used as the source of lime in agriculture, aquatics, and in power plants. Evaporite deposits are mined for gypsum (used in drywall), halite (road and table salt), and potash (potassium salts used in fertilizer).

General Mining Law of 1872

The General Mining Law was passed to encourage prospectors to use federal land and to promote commerce. The law allows miners to stake claims on federal land and take any minerals they find. They can purchase land for $5 an acre or less, and then they own the land as private property. For little money, companies have purchased land that contains billions of dollars worth of minerals. Land also may be resold for development. Mining companies can take any ore that they wish without extra payments to the government. Worse still, they can deduct a depletion allowance from their taxes, which lowers their taxes as the resource declines. Many people suggest that mining companies pay royalties on the ore extracted from these mines to reflect the profit being made by the companies. They also want legislation requiring remediation of mined lands.

> Take Note: You must be able to delineate the damage to an ecosystem incurred when mineral extraction takes place. Also, be able to describe environmental damage associated not only with mining of minerals, but of smelting the ores to remove the desired metals. Free response questions often ask the origin of a specific pollutant, and students frequently forget that smelting is the origin of heavy metal and sulfur dioxide emissions.

Environmental Impacts of Resource Extraction

Mining results in the removal of land surface to get to the mineral desired. This destruction of land causes habitat loss for wildlife and deforestation and induces erosion. Additionally, due to the machinery involved in extraction, air pollution may be high in these areas, including sulfur dioxide and particulate pollution. Water pollution is common. Water may be contaminated by the sediments from erosion. Many minerals are frequently found in conjunction with sulfide ores, which may form sulfuric acid when exposed to water. The release of sulfuric acid from mines is called acid mine drainage. The drainage can acidify streams, resulting in death of pH sensitive species, which decreases stream biodiversity. There is also an increased toxicity due to aluminum solubility when streams are acidified. Frequently cyanide, mercury, or other toxins are used to extract minerals from mines.

The main types of mines are placer mines, open pit mines, strip mines, and subsurface mines. Placer mines remove valuable minerals such as gold, diamonds, and coal from streams. They contaminate surface water ecosystems because they use water cannons to remove minerals from hillsides. Open pit mines, or quarries, are used for sand, gravel, copper, and a variety of other materials. These mines frequently fill with water after they have been abandoned, resulting in acidification or concentrated heavy metals depending upon the original type of mine. Strip mines, or surface mines, are frequently used to remove coal from the western United States. These mines remove the topsoil from a strip of land, then remove the surface rock, or overburden. The mineral is then extracted, and the overburden is replaced in the mines in long ridges called spoil banks. These banks are susceptible to erosion and weathering. There is no topsoil, so vegetation is slow to recover in these areas. In 1977, Congress passed the Surface Mining Control and Reclamation Act (SMCRA), which requires that mined areas be returned to their approximate original topography and replanted. Subsurface mines have been used to harvest tin, lead, copper, and coal. Underground mines frequently collapse or have explosions from dust or natural gas leaks. Some coal mines catch on fire and remain smoldering for years. One fire in China has been burning for 400 years, and one in Pennsylvania has been burning for over 40 years. Subsurface mines may strike groundwater and contaminate it with heavy metals.

A new type of mining known as mountaintop removal has developed in the Appalachians. This mining involves removing the overburden from the top of a seam of coal and placing it in a nearby valley. The coal is then extracted from the mine. The entire topography of an area is altered when mountaintop removal is used. Miners apply for a variance to the SMCRA to be able to carry out mountaintop removal.

Smelting is the heating of an ore to extract the desired metal. Tremendous amounts of air pollution are released during the smelting process. The fossil fuel used to melt the ore will produce its own air pollutants, and the heating of the ore typically releases large amounts of sulfur dioxide. The city of Ducktown, Tennessee, is notorious for the copper smelting that occurred in the area, which resulted in ruination of the entire ecosystem due to acid deposition when the sulfur dioxide emitted from the process acidified the soil.   The area has been replanted and treated to improve the pH of the soil to a more neutral level. Arsenic and lead may be present in ores and may be released during the smelting process.

Heap leach extraction is the process of spraying ore with a cyanide solution to dissolve gold or silver, then removed in an electrolysis process. In 2000, this method resulted in the contamination of the Danube River with cyanide and heavy metals from a leaking wastewater lagoon at a smelter in Romania. The cyanide killed hundreds of fish as it progressed down the Danube through Hungary and Yugoslavia. The spill also impacted species such as birds, which feed upon fish.

Conserving Geological Resources

One of the easiest ways to preserve geological resources is by recycling metals. To create a market for recycled material, it must be economically profitable to recycle. One of the most successful recycling programs is the recycling of steel. Nearly all steel in the United States contains at least 25 percent recycled steel, and some contain 100 percent recycled steel. Much of this steel comes from steel cans, automobiles, buildings and bridges, and appliances. Mining and smelting iron to obtain steel consumes far more energy than melting old steel and forming new steel cans or girders. Aluminum, derived from an ore called bauxite, must be mined and extracted from the ore in a complicated process. By recycling aluminum in cans, the costs associated with the extraction of virgin ore are dramatically lowered. Aluminum is also easy to recycle and can easily be melted and remade into a can almost immediately. Other metals that may be recycled include lead (auto batteries), platinum (catalytic converters), gold, silver, copper (pipes and electrical wiring), and iron.

Natural resources may also be saved by developing new materials to replace those formed from metals. For example, polyvinyl chloride (PVC) pipes are now used in plumbing to replace copper, lead, and steel pipes. Ceramics, alloys (titanium mixed with steel), and aluminum are used to replace metals in the engine and frames of automobiles. The copper and aluminum wire used in years past for wiring electronics is replaced by glass cables.

Geologic Hazards

Volcanoes typically occur at convergent or divergent plate boundaries. An exception to this volcano forming geology is the Hawaiian Islands, formed by a mantle plume from a hot spot in the middle of an oceanic plate. There are currently 550 active volcanoes in the world and countless dormant volcanoes. Volcanic eruptions cannot be predicted, but a volcano may give several warning signs that it is about to erupt, including increased seismic activity; gas emissions; and pyroclastic debris, including ash. Humans are killed not only by lava but also by the large amounts of ash and toxic gases emitted during an eruption. Gases emitted during an eruption include water vapor; carbon dioxide; nitrogen; carbon monoxide; and sulfur gases, including sulfur dioxide and hydrogen sulfide. Water vapor comprises up to 80 percent of the gases released during an eruption. Mudslides are often caused by volcanic eruptions and contribute to human deaths. Volcanic dust tends to block sunlight, resulting in cooler regional temperatures. Additionally, the sulfur dioxide emitted combines with atmospheric water to create a sulfur haze, which reflects sunlight and contributes to global cooling. When Mount Pinatubo in the Philippines erupted in 1991, the sulfur haze that formed cooled the earth by 1°C.

Earthquakes are movements that occur along fault lines where plate boundaries meet. The area under the earth that is the origin of the earthquake is called the focus, and the site on the surface of the land that is the origin is known as the epicenter. Earthquakes tend to have a series of aftershocks, or the smaller quakes that occur after a quake. Earthquakes are particularly dangerous because they cannot be predicted. People that die during an earthquake are usually killed by substandard construction. Buildings constructed in earthquake prone areas should adhere to stringent standards for construction, so that not only are the buildings maintained, but the loss of life is minimized. Earthquakes under water, landslides into the ocean, or volcanic eruptions can cause a tsunami. The tsunami of 2004 that occurred in the Indian Ocean killed more than 300,000 people. The Ring of Fire in the Pacific Ocean is particularly vulnerable to tsunami formation because it is such a geologically active area. The United States has deployed tsunami detection buoys to allow for early notification that a tsunami is on its way. A worldwide detection system is to be established by 2015 after countries pledged to prevent deaths similar to those that occurred in Southeast Asia in 2004.

Floods are the leading cause of death by natural disaster, but storms such as hurricanes or tropical cyclones cause the most property damage.

Mass wasting is the movement downhill of large amounts of earth, rocks, or ice at one time. The material can move quickly, as in an avalanche or landslide, or slowly, as in a mud flow. Mass wasting can cause enormous amounts of property damage when homes or businesses are built upon unstable ground.

Chapter 14 Questions

1. The most prevalent element in the earth's crust is
a. nitrogen.   b. potassium.   c. sulfur.   d. hydrogen.   e. oxygen.

2. The most prevalent element in the earth's core is
a. lead.   b. nickel.   c. mercury.   d. iron.   e. uranium

3. The most common rock on earth is
a. sedimentary.   b. igneous.   c. metamorphic.   d. limestone.   e. marble.

4. Which of the following is not a metamorphic rock?
a. schist   b. marble   c. limestone   d. anthracite   e. slate

5. Which of the following is an example of chemical weathering?
a. a rock fractured by the freezing of water in the cracks of the rock
b. a rock in the desert expands when heated during the day then contracts at night, which breaks off pieces of rock
c. rushing water smooths rocks in a streambed
d. rocks in a glacier scrape the underlying bedrock, causing fractures
e. a rock loses particles upon exposure to rain, which dissolves portions of the rock

6. Which of the following statements is true regarding volcanic eruptions?
a. Lava is always emitted during a volcanic eruption.
b. Water vapor makes up the greatest proportion of gases during a volcanic eruption.
c. Most people killed during an eruption are killed by the lava flow.
d. No volcanoes have erupted in the continental United States since 1920.
e. Volcanic eruptions contribute to global warming.

7. All of the following are environmental impacts of mining except
a. depletion of ore resources.   b. increased erosion.
c. deforestation.   d. loss of biodiversity.
e. acid mine drainage.

8. The geologic hazard that results in the most human deaths is
a. volcanoes.   b. earthquakes.   c. floods.   d. mass wasting.   e. landslides.

9. The San Andreas Fault is found at a
a. transform boundary.   b. diverging plate boundary.   c. converging plate boundary.
d. subduction zone.   e. oceanic ridge system.

10. Which of the following volcanoes does not occur at a plate boundary?
a. Mt. Pinatubo, Philippines   b. Mt. St. Helens, Washington
c. Kilauea, Hawaii   d. Cotopaxi, Ecuador
e. Mt. Etna, Italy

Chapter 14 Answers

1. e. Oxygen is the most prevalent element in the earth's crust. It is found in many types of rocks.
2. d. Iron is the most prevalent element in the earth's core.
3. b. Igneous is the most common rock on earth. Sedimentary rock is common on land, but most of the ocean floor is igneous. Limestone is a sedimentary rock, and marble is metamorphic.
4. c. Limestone is a sedimentary rock.
5. e. If rain dissolves portions of a rock, then the rock is undergoing chemical weathering. The other examples are mechanical weathering.
6. b. Water vapor makes up the greatest proportion of gases during a volcanic eruption. Lava is not always emitted during a volcanic eruption, and humans are frequently killed by the gases and ash. Mount St. Helens in Washington erupted in 1980. Volcanic eruptions usually result in global cooling, due to the ash and sulfur aerosol decreasing sunlight to the planet.
7. a. Depletion of ore resources is an economic impact of mining.
8. c. Floods are the geologic hazard that results in the most human deaths.
9. a. The San Andreas Fault is found at a transform boundary.
10. c. Kilauea does not occur at a plate boundary. It occurs at a mantle plume in an ocean plate.

Chapter 15 - Air, Weather, and Climate

Key Terms

aerosols	greenhouse effect	stratosphere
albedo	hurricane	tornado
climate	jet stream	tropical cyclone
cold front	Kyoto Protocol	troposphere
convection currents	La Niña	warm front
Coriolis effect	latent heat	weather
downburats	Milankovitch cycles	
El Niño Southern	monsoon	
Oscillation Event	ozone	

Skills

1. Characterize human contributions to global climate change. Outline the effects climate change is having on ecosystems.
2. Discuss the composition and layers of the atmosphere.
3. Explain how climate is generated by the prevailing winds, ocean currents, jet streams, and continental geography.
4. Identify and describe weather hazards.
5. Examine El Niño and La Niña impacts on global weather. Identify the physical phenomena that generate an El Niño Southern Oscillation Event.

Take Note: You are expected to understand basic information regarding the earth's atmosphere, weather, and climate. The weather determines the biomes that are present throughout the earth. Released multiple-choice questions have examined convection currents, fronts, and weather hazards.

The Atmosphere

The atmosphere has four layers. Closest to the surface of the earth is the troposphere, followed by the stratosphere, mesosphere, and thermosphere. Air circulates in the troposphere because warmer air rises and colder more dense air sinks, forming convection cells. Gravity causes the troposphere to be the densest of all of the layers. The troposphere is comprised of 78 percent N_2, 21 percent O_2, a little less than 1 percent argon, and 0.035 percent CO_2. Water vapor varies regionally from 0–4 percent. Aerosols, created by suspended solid and liquid particles, are important in reflecting sunlight and in serving as condensation nuclei for cloud formation.

The stratosphere is less dense than the troposphere, and contains a layer of ozone, responsible for absorbing UV-B radiation. The ultraviolet radiation would damage living tissues, so the stratospheric ozone layer is responsible for allowing the evolution of and continued survival of life on land. Stratospheric ozone is decreasing due to human pollution, primarily from released chlorofluorocarbons (CFCs). The increased levels of UV-B are likely to increase human skin cancers and cataracts; damage plant tissues, which will lower crop productivity; and decrease biodiversity due to loss of UV sensitive species. The mesosphere forms between the stratosphere and the thermosphere. The thermosphere is heated by solar and cosmic radiation. The northern and southern lights, aurora borealis, and aurora australis, are caused by the glowing of ions in the lower thermosphere.

Weather is the daily changes in precipitation and temperature. An area's climate is the long-term weather patterns observed in the area.

The Sun and Convection Cells

The sun's energy is focused more directly on the equator than the poles, which makes the equator hotter than the poles. One-fourth of the sun's energy is reflected back into space by the atmosphere and another quarter is absorbed by atmospheric gases. Some of the energy that reaches earth is reflected back into space due to the albedo effect of light surfaces, such as snow, sand, and ice. Much of the remainder of the energy is absorbed by darker surfaces, which heats the earth. Some of the absorbed energy (about 1 percent) is used for photosynthesis and the rest serves to increase evaporation of water. The energy from the sun is ultimately transformed into the less useful, more stable thermal energy, as expected based upon the second law of thermodynamics. Most of this energy, now longer wavelength infrared energy, is trapped in the earth's atmosphere for the most part, which has the net effect of warming the earth. The phenomenon is known as the greenhouse effect, because the infrared energy is kept from exiting the earth's atmosphere by gases, much like the heat in a greenhouse is prevented from leaving due to the glass walls. This effect is exacerbated by human activities such as deforestation and fossil fuel combustion, resulting in a greater heating of the earth's surface. The sun not only heats the earth, it also serves to drive the hydrologic cycle by promoting evaporation. Water contains stored energy, known as latent heat. The evaporated water then moves via air currents to other parts of the globe.

The uneven heating of the earth's surface causes the warmer air to rise and the cooler air to sink, creating convection cells. All of the large convection cells on earth create the global circulation patterns for air. The air patterns contribute to the global circulation patterns for water.

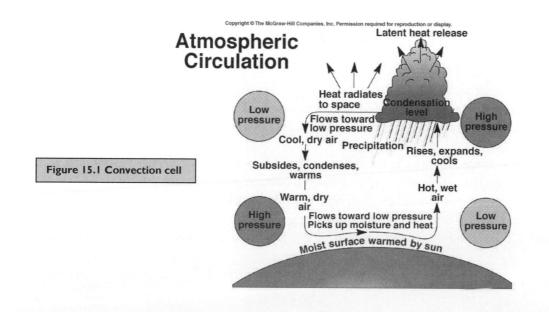

Figure 15.1 Convection cell

Precipitation

As the water in the atmosphere condenses when the air cools, the water droplets form clouds around condensation nuclei. Areas with high levels of precipitation have atmospheric circulation that contributes to the rising and cooling of the warm moisture laden surface air. These areas include regions with colliding air masses, the windward side of mountains, and around the equator. Sinking dry air occurs at 30° north and south of the equator and on the leeward side of mountains. In these regions, deserts form. The formation of deserts on the dry, leeward side of mountains is attributed to the rain shadow effect. The clouds release their moisture as they rise, so only the windward side of mountains receive precipitation, thus creating a "shadow" that lacks rain on the opposite side of the mountain.

15.2 Rain shadow effect

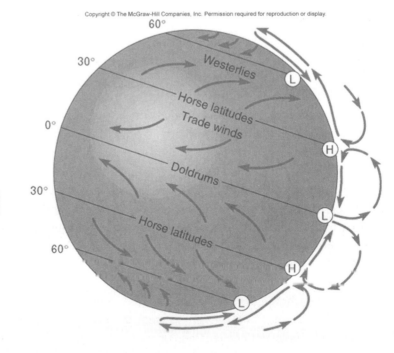

Figure 15.3 Global circulation patterns

Wind and Ocean Currents

Due to the eastward rotation of the earth, winds in the northern hemisphere deflect to the right (clockwise) and winds in the southern hemisphere deflect to the left (counterclockwise). This deflection is known as the Coriolis effect. This effect creates the wind patterns observed in figure 15.3. Overlying wind patterns known as jet streams also circulate the earth in the troposphere. These jet streams cannot be felt on the ground, but they have a profound impact on weather. These winds are not stationary and dip down into lower or higher latitudes.

The surface ocean currents also move in a pattern similar to the winds, creating deep ocean currents. Ocean currents are also influenced by the circulation of the wind. Differing density of water, due to changes in temperature and salinity, also play a role in ocean circulation. The cycling currents formed in the large oceans of earth are known as gyres, and they distribute heat from the lower to the upper latitudes. For example, the Gulf Stream, which flows from the Caribbean to Europe, creates a warmer microclimate in England than would be expected based upon England's latitude.

Seasonal winds and rains, known as monsoons, form in several tropical regions. These rains occur as hot air over the ocean is blown across continents, where the air rises, creating precipitation. Subtropical or tropical areas typically experience dry and wet seasons.

Fronts

A front is a boundary between two air masses of different density and temperatures. Cold fronts occur when a cold dense mass of air advances on a warmer mass. The cold mass sinks under the warm mass, creating strong thunderstorms and rain. Clouds associated with a cold front are called thunderheads, and they are shaped like an anvil. Warm fronts occur when a warm air mass advances on a cooler air mass. The warm air rises, resulting in drizzle and clouds throughout several levels of the troposphere.

Severe Weather

Severe weather includes tropical cyclones and tornadoes. Tropical cyclones, called hurricanes in the Atlantic and eastern Pacific and typhoons in the western Pacific, are generated when a low-pressure system develops over warm ocean water. The latent energy released by condensation causes convection cells to increase their circulation rate.   The spinning storm has high wind speeds that increase in intensity as the center of the storm is reached. The damage from a hurricane is not only due to the high wind speeds, but the water pushed ashore by the rapidly moving winds. This wall of water, called the storm surge, is extremely dangerous due to the rapid rate at which it rises on land. Hurricanes release large amounts of precipitation and spawn tornadoes as they move over land. Hurricane Katrina in 2005 had a storm surge of 29 feet and had 145 mile an hour winds. The coastal devastation is still being evaluated, but entire towns were destroyed.

Tornadoes occur when a strong, dry cold front collides with warm, humid air. The greater the temperature differences of the air masses, the more severe the tornado. As the warm air rises over the cold air, internal convection currents strengthen, forming the tornado. When the storm touches the ground the devastation is tremendous. If the spinning action of the tornado does not begin, a downburst may occur. Downbursts are powerful downdrafts of air that can generate extremely high winds and property damage. Hail is frequently present during tornadoes because the water particles are drawn up the tornado then sink again. The cycle repeats until the hail is too heavy to fly upwards again.

Climate and Climate Change

Through study of polar ice sheets, a correlation has been established between carbon dioxide levels and mean global temperatures. When climate change is slow, populations can undergo natural selection to become better suited to the new climate, or they can migrate to a more suitable climate. If the climate change occurs rapidly, organisms will die out because they are not adapted to the new climate. An example of this mass death due to climate change occurred 65 million years ago at the end of the Cretaceous period. This mass extinction due to global cooling killed all of the dinosaurs, which opened up numerous niches to allow adaptive radiation of mammals.

There are thought to be several explanations for major climate shifts. There are changes in the sun's energy output over time. There are also shifts in the moon's orbit, which alter tides and circulation, thus affecting climate. Milankovitch cycles, periodic shifts in the earth's tilt and orbit, also explain extreme climate shifts. Volcanoes that produce massive amounts of ash and sulfur dioxide would cause global temperatures to drop quickly.

> **Take Note:** You must be familiar with the changes in global climate associated with an El Niño Southern Oscillation event (ENSO). One essay question asked about the cause of an ENSO and the relationship of disease transmission to the altered climate.

El Niño Southern Oscillation Event

The El Niño Southern Oscillation event (ENSO) is comprised of an El Niño and the intervening years called La Niña. The event is named El Niño because Peruvian fishermen noted that the phenomenon tended to begin in December around Christmas, so they named it *el niño* for little boy, or Jesus. The root cause of an El Niño is a slowing of the equatorial trade winds that holds warm water in the Pacific Ocean close to Indonesia. The slowing of the trade winds allows the warm waters to move across the Pacific to South America. This slowing of the equatorial winds occurs roughly every three to five years and last about a year. This warm water inhibits the upwelling that occurs along the continental edge of South America. Upwelling brings the cool, nutrient-rich waters from the ocean floor up to the surface. The upwelling provides nutrients that support algal growth that then serves as a food supply for the anchovies that are so important to the South American people and wildlife. Of late, these events are stronger and more irregular than in the past. Although El Niño takes place in the Pacific Ocean, there are effects felt globally, as well. In the western United States, which borders the Pacific, there is greater moisture in the air, resulting in more damaging storms and heavy rains. Many mudslides occur in California during El Niño events. The northwestern United States tends to be sunny instead of its typical rain. The likelihood of hurricanes in the Atlantic are diminished because the jet stream usually in Canada drops farther south and inhibits hurricane formation. Australia and Indonesia experience severe drought.

- 140 -

> **Take Note:** All students enrolled in an AP environmental course are expected to have a thorough understanding of global climate change; the origins of the gases involved; methods to reduce the greenhouse effect; and the expected impacts of the climate change, including human health impacts and environmental impacts. The 2006 AP exam had one entire essay about global warming. It is imperative that you are familiar with the following material.

Global Warming

Humans are playing a role in the increase in global temperatures. Greenhouse gases include CO_2, methane, water vapor, N_2O, and CFCs. CO_2 levels are so low in the atmosphere, any slight increase has a profound impact on the global climate. The oscillations in the line shown in figure 15.4 can be attributed to the seasonal changes in CO_2 in the atmosphere due to increased photosynthesis in the summer months in the northern hemisphere.

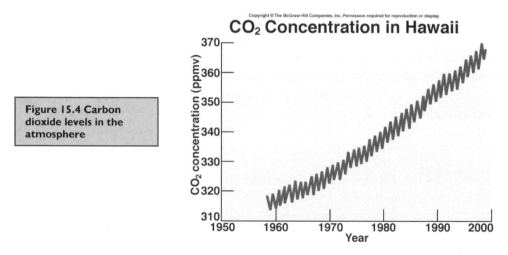

Figure 15.4 Carbon dioxide levels in the atmosphere

Anthropogenic carbon dioxide is produced during combustion of fossil fuels, biomass, and refuse. Deforestation contributes to high levels of CO_2 in the atmosphere because the forests serve as carbon sinks and they can no longer remove CO_2 from the atmosphere. Methane is 20 times more efficient at holding heat in the atmosphere than is CO_2. Methane is released during decomposition in landfills and during recovery and processing of petroleum. CFCs also absorb infrared energy. Nitrous oxide comes primarily from livestock feedlots and combustion of organic material.

Sulfur dioxide acts in opposition to the greenhouse gases. It forms a sulfur aerosol that reflects sunlight and results in global cooling.

In 1988 the Intergovernmental Panel on Climate Change was formed to address global temperature increases. The panel concluded that humans are playing a role in climate change. Climate change will have a profound impact on glaciers. As the earth warms, continental glaciers will melt, causing rising sea levels. The rising sea levels will dramatically impact human settlements in coastal areas. As the glaciers melt, the albedo effect will decrease, resulting in less sunlight reflected into space, which will further increase the global temperature. Krill, an important food for numerous marine species, will also decline as a result of the ice sheets melting. The loss of this food source will have a dramatic impact for the organisms that feed upon them. The organisms will have to migrate or face extinction. Droughts are becoming more frequent and widespread. Amphibians are vanishing in large numbers throughout the world, and their disappearance has been linked to the increasing global temperatures. Infectious disease is likely to increase in many areas as insects move to new regions due to the climate change. Mosquitoborne illnesses such as malaria, West Nile Virus, and dengue are springing up in new locales. Although increased CO_2 will likely have a negative impact on animal species, some plants will thrive with additional CO_2 available for photosynthesis.

Kyoto Protocol

At the 1992 Rio Earth Summit, the Framework Convention on Climate Change created a goal of stabilizing greenhouse gases to reduce global warming. A subsequent conference in 1997 was held in Kyoto, Japan. At this conference 160 countries agreed to reduce greenhouse gas levels to below the 1990 levels by 2012. This agreement is known as the Kyoto Protocol and has now been signed by all industrialized countries except Monaco, Liechtenstein, Australia, and the United States. Poor countries such as China and India are exempt from the reduction agreement as they are encouraging development to improve their standard of living. The United States was very influential in generating the Kyoto agreement under President Clinton's administration, but President Bush has refused to sign it. As a result, American companies that have production facilities in other nations may face sanctions and other penalties due to the president's refusal to sign the treaty. The agreement establishes pollution credits that can be bought or sold.

Reducing Greenhouse Gas Emissions

The easiest way to reduce greenhouse gas emissions is to promote energy conservation and efficiency. Natural gas can also be used to generate electricity instead of oil or coal because it releases less CO_2. Nuclear power and the alternative energy resources of wind, solar, geothermal, and hydroelectric reduce greenhouse gas emissions. Mass transit, carpooling, and bicycling will decrease CO_2 from gasoline combustion. Planting trees creates forests that serve as carbon sinks. CO_2 can be piped underground for long-term storage.

Chapter 15 Questions

Use the following choices for questions 1–4.

        a. methane
        b. nitrous oxide
        c. carbon dioxide
        d. CFCs
        e. sulfur dioxide

1. greenhouse gas emitted by feedlots and decomposition in landfills
2. greenhouse gas increased due to deforestation
3. gas that contributes to global cooling
4. greenhouse gas that arises from fossil fuel combustion and from agricultural fields after fertilizer is applied

5. Ozone in the stratosphere protects earth from
a. cosmic rays.   b. meteorites.   c. ultraviolet radiation.   d. infrared radiation.
e. microwaves.

6. The gas not found in earth's early atmosphere was
a. oxygen.   b. hydrogen.   c. nitrogen. d. methane. e. carbon dioxide.

7. Which of the following exhibits the least albedo?
a. snow   b. icepack   c. sand   d. water e. forests

8. An El Niño will result in all of the following environmental effects except
a. increased hurricanes in the southern United States.
b. increased death of sea lions in the Galapagos Islands.
c. decreased precipitation in Australia.
d. increased mosquitoborne disease in tropical areas.
e. decreased rain in the northwestern United States.

9. Which of the following weather events results in the least amount of precipitation?
a. hurricane   b. typhoon c. cold front   d. warm front   e. monsoon

10. All of the following increase carbon dioxide emissions except
a. animal feedlots.   b. deforestation.   c. burning fossil fuels.
d. incineration.   e. combustion of biomass.

Chapter 15 Answers

1. a. Methane is a greenhouse gas emitted by feedlots and decomposition in landfills.
2. c. Carbon dioxide is a greenhouse gas increased due to deforestation.
3. e. Sulfur dioxide contributes to global cooling by creating a sulfur aerosol.
4. b. Nitrous oxide is the greenhouse gas that arises from fossil fuel combustion and from agricultural fields after fertilizer is applied.
5. c. Ozone in the stratosphere protects the earth from ultraviolet radiation.
6. a. Oxygen was not found in the earth's early atmosphere. It did not arise until photosynthetic organisms evolved.
7. e. Forests are dark in color, thus they exhibit the least albedo.
8. a. An El Niño will result in all of the following environmental effects except increased hurricanes in the southern United States. The hurricane number declines in an El Niño year.
9. d. A warm front produces the least amount of precipitation.
10. a. Animal feedlots release nitrogen oxide and methane, but not carbon dioxide.

Chapter 16 – Air Pollution

Key Terms

acid deposition	chronic obstructive	primary pollutant
aerosol	pulmonary disease	secondary pollutant
aesthetic degradation	criteria pollutants	stratospheric ozone
ambient air	fugitive emissions	sulfur dioxide
bronchitis	hazardous air pollutants	temperature inversions
carbon dioxide	nitrogen oxides	Toxic Release Inventory
carbon monoxide	ozone	volatile organic compounds
chloroflourocarbons	particulate material	

Skills

1. Differentiate between the major types of air pollutants. Address sources, environmental impacts, and human health effects of these pollutants.
2. Compare the criteria pollutants to the unconventional air pollutants.
3. Characterize the indoor air pollutants, including their sources, types of buildings affected, and human health effects.
4. Explain what is meant by a sick building, and delineate the criteria required for a building to be considered sick.
5. Differentiate between stratospheric and tropospheric ozone.
6. Propose specific mechanisms to decrease each of the indoor and outdoor air pollutants.
7. Appraise the effectiveness of the Clean Air Act.
8. Recall the chemical reactions involved in contamination of the atmosphere with pollutants.

Take Note: It is important that you understand the impacts of the major air pollutants, including their human health effects and environmental impacts. You should also know the origin of these pollutants and methods to reduce the pollutant. Previous AP essay questions have addressed ozone, lead, mercury, particulates, carbon dioxide, and carbon monoxide. In addition, one essay question was entirely about indoor air pollutants. Multiple-choice questions will also reference the pollutants.

Natural Sources of Air Pollution

Volcanoes are a natural source of particulates, sulfur dioxide, carbon oxides, hydrogen sulfide, and other noxious air pollutants. Forests release large amounts of volatile organic compounds that form ozone upon exposure to sunlight. Sea spray releases sulfur compounds into the atmosphere, as does decaying vegetation. Dust creates particulate pollution. Natural decomposition releases tremendous amounts of the greenhouse gas methane.

Anthropogenic Air Pollutants

Primary air pollutants are chemicals that are considered pollutants when they are released. A secondary air pollutant forms when exposure to another factor causes the chemical to be toxic. Many times secondary pollutants are a result of a photochemical reaction or a reaction with water. For example, photochemical oxidants like ozone and atmospheric acids like nitric or sulfuric acids are secondary air pollutants. Fugitive emissions are emissions that do not arise from a smokestack. They may be from leaking pipes and valves or may be dust from erosion, mining, or construction.

According to the Environmental Protection Agency, the United States has deemed six pollutants as criteria air pollutants. These pollutants have legal limits, or national ambient air quality standards (NAAQS), which control their release into the environment. The pollutants are ozone, nitrogen dioxide, particulate matter, sulfur dioxide, carbon monoxide, and lead. Other air pollutants that must be considered include volatile organic compounds (VOCs), carbon dioxide, and mercury.

Air Pollution and Human Health

Often, preexisting lung diseases are exacerbated upon exposure to air pollutants. Bronchitis, or inflammation of the bronchi and bronchioles, results in constricted airways. Chronic obstructive pulmonary disease (COPD), or emphysema, is the breakdown of the alveoli in the lungs. As a result, the surface area for gas exchange is greatly lowered, resulting in decreased absorption of oxygen. Smoking cigarettes, or exposure to environmental tobacco smoke, is the leading cause of COPD and preventable death in the world. Each of the pollutants addressed in this chapter has specific health impacts that will be addressed as the pollutant is discussed.

Sulfur dioxide

Sulfur compounds enter the atmosphere from a variety of natural and anthropogenic sources. Natural sources include sea spray, volcanic gases, and biogenic sulfur compounds from bacterial decomposition. Anthropogenic sources include smelting, coal combustion, and petroleum distillation. Sulfur dioxide can directly damage the tissue of living organisms. Once it is in the atmosphere it readily oxidizes to form sulfur trioxide (SO_3). The SO_3 reacts with water vapor (H_2O) in the atmosphere to form sulfuric acid (H_2SO_4), a secondary air pollutant. The sulfuric acid, in wet or dry form, is one of the primary acids involved in acid deposition. Sulfate particles create the sulfur haze that reflects sunlight, causing a global cooling. The particles reduce visibility, as well.

Nitrogen Dioxide

Nitrogen oxides are released when fossil fuels are burned. The first pollutant released is nitric oxide (NO), which readily oxidizes to form nitrogen dioxide (NO_2). When nitrogen dioxide reacts with water vapor it forms nitric acid (HNO_3), a secondary air pollutant that is the other major acid involved in acid deposition. NO_x refers to all of the nitrogen and oxygen compounds that are air pollutants. Another NO_x is N_2O, a greenhouse gas, which arises from feedlots and use of fertilizers.

An important effect of nitrate release in the atmosphere is contamination of aquatic systems that results in eutrophication. It is also suspected that these excess atmospheric nitrates may be enriching soil and promoting the growth of weed species.

Ozone

Photochemical oxidants are created when primary pollutants are acted upon by photons of light. When nitrogen dioxide loses an oxygen atom when exposed to a photon, that free radical oxygen combines with a molecule of O_2 to create ozone (O_3). Ozone is a powerful oxidant that damages vegetation, building materials, and mucus membranes. In humans, ozone irritates the respiratory system resulting in discomfort, coughing, and throat irritation, and it exacerbates lung disease. Ozone is a major component of photochemical smog.

Particulate Matter

An aerosol is any solid or liquid suspended in gas. All atmospheric aerosols, regardless of state, are considered particulate matter (PM). Examples of PM include acid droplets, ash, soot, dust, microscopic biological particles, and smoke. Erosion and desertification dramatically increase the PM in the atmosphere. PM also arises from combustion of fossil fuels or incineration. Power plants frequently use controls for particulate emissions, so most of the particulate pollution in the United States is from diesel engines. Some municipalities in the United States are banning wood-burning stoves and fireplaces due to the particulate pollution. The primary environmental impacts of PM are to decrease visibility and coat plants and structures with deposits. Any small particle, less than $2.5 \mu m$ in diameter, is particularly dangerous to human health because smaller particles are more likely to be deeply inhaled and affect lung tissue. Larger particles tend to settle out of the air or be trapped in the hair and mucus lining human respiratory systems. Particulates irritate the respiratory tract, leading to coughing and difficulty breathing. Exposure to PM also exacerbates lung and heart disease and are linked to heart attacks, asthma, bronchitis, lung cancer, and even abnormal fetal development.

Carbon Monoxide

Carbon monoxide, CO, is a colorless, odorless gas produced during the incomplete combustion of fossil fuels or biomass. Most of the CO in the United States is released by internal combustion engines. In other countries, fires to clear vegetation and cooking fires also release large amounts of CO. Its primary deleterious effect is on human health. CO binds irreversibly to hemoglobin with more affinity than does oxygen, resulting in suffocation. Acute symptoms of CO exposure are fatigue, impaired vision, headache, dizziness, nausea, and confusion. High concentrations of CO are lethal.

Lead

The primary atmospheric source of lead in the United States was leaded gasoline, which was banned, resulting in a 90 percent decline in the blood levels of lead in American children. Lead is a neurotoxin that binds enzymes and inactivates them. Lead bioaccumulates in bone, and this chronic exposure to an endogenous toxin induces developmental retardation, impaired IQ, attention deficits, hyperactivity and learning disorders, and aggression. Lead also binds hemoglobin and reduces the ability of red blood cells to carry oxygen.

Volatile Organic Compounds

Volatile organic compounds enter the atmosphere as a gas. Most VOCs are produced naturally by forests. Biogenic methane also contributes to the VOCs in the atmosphere. Usually these VOCs are degraded into CO and CO_2 in the atmosphere. Synthetic VOCs released into the atmosphere include formaldehyde, benzene, toluene, vinyl chloride, phenols, and chlorotorm. The origin of these pollutants ranges from combustion of fossil fuels, refineries, and chemical plants. VOCs, like NO_x, contribute to the formation of photochemical smog.

Photochemical Smog

Photochemical smog, or brown air smog, occurs in warm sunny areas with large numbers of automobiles. The chemical reactions involved in the formation of photochemical smog are shown in figure 16.1. Ozone and the other photochemical oxidants irritate mucous membranes in animals and directly damage plant tissues. Smog also reduces visibility due to the haze that it creates.

Figure 16.1 Chemical reactions involved in the generation of photochemical smog

1. NO + VOC $\longrightarrow$ NO_2 (nitrogen dioxide)
2. NO_2 + UV $\longrightarrow$ NO + O (nitric oxide + atomic oxygen)
3. O + O_2 $\longrightarrow$ O_3 (ozone)
4. NO_2 + VOC $\longrightarrow$ PAN, etc. (peroxyacetyl nitrate)

Net results:

NO + VOC + O_2 + UV $\longrightarrow$ O_3, PAN, and other oxidants

Carbon Dioxide

The major natural source of carbon dioxide, CO_2, in the atmosphere is cellular respiration. The level of CO_2 in the atmosphere is increasing due to deforestation, burning of biomass, and combustion of fossil fuels. Carbon dioxide is a greenhouse gas, and it will form carbonic acid (H_2CO_3) when it reacts with water vapor in the atmosphere and thus contributes to acid deposition.

Mercury

Mercury is released naturally by volcanoes, weathering of rock, and evaporation of sea water. Mercury contamination arises anthropogenically from coal combustion, incineration, and smelting. Mercury in the atmosphere enters surface water via precipitation, and the inorganic mercury is methylated by bacteria to create the most toxic form, called methyl mercury. Inorganic mercury salts may be in fungicides and disinfectants. Human exposure is from eating contaminated fish and shellfish. Sharks, swordfish, kingfish, and tilefish contain high levels of methyl mercury because the metal is readily bioaccumulated and biomagnified. The acute effects of mercury toxicity are difficulty walking, loss of coordination, difficulty swallowing, and tremors. Chronic effects are due to the mutagenic, teratogenic, and carcinogenic effects of the metal. Additional effects include hallucinations, psychosis, and irreversible brain damage. Fetal exposure results in mental retardation, attention disorders, seizures, and blindness. The blood level considered to be safe is less than 5.8 µg/l. Mercury emission from power plants is regulated with pollution credits rather than strict emission controls.

Other Metals

Other atmospheric pollutants include arsenic, nickel, beryllium, cadmium, thallium, uranium, cesium, and plutonium. Arsenic is released from smelters and coal combustion. Acute exposure induces anemia and nausea in humans. Chronic exposure induces chronic fatigue, gastrointestinal disease, anemia, and eventually death. Arsenic is linked to numerous types of cancers and birth defects.

Hazardous Air Pollutants

Hazardous air pollutants (HAPs) are those pollutants that cause cancer, birth defects, mutation, or that are neurotoxic, immunotoxic, or endocrine disrupters. The EPA requires that companies relay to the public their Toxic Release Inventory, a record of any toxic materials released over the minimum amount in a year.

Noise Pollution

Noise pollution is any unwanted or undesirable anthropogenically created sound. Noise pollution increases aggression and blood pressure. Examples of noise pollution would be congested traffic, loud music, or airport traffic. Chronic exposure to loud sound may induce deafness. Noise pollution not only reduces the quality of life for human beings, but it may interfere with animal behavior such as migration, courtships, or circadian rhythms.

Geography Can Exacerbate Air Pollution

Geography plays an important role in the impact of air pollution. Temperature inversions typically occur in a valley. The air closest to the ground, filled with pollutants, is prevented from rising by a layer of colder air. The pollutants thus remain concentrated, and the human health impacts of the pollutants are far greater. Examples of cities that suffer from temperature inversions are Mexico City and Los Angeles, both surrounded by mountains. Photochemical smog becomes worse throughout the day as more NO_x and VOCs are added through vehicle traffic and industrial processes. The pollutants cannot escape due to the night cap of cool air and become worse over time.

Cities also create urban heat islands, due to the lack of cooling green vegetation and the microclimate that the vegetation creates. The heat is also generated because of increased precipitation runoff due to lack of permeable surfaces. The heat holds in pollutants, particularly dust and particulates, creating dust domes.

Air pollutants from one area can enter atmospheric circulation and fall in the convection cells in anther area. Therefore many air pollutants tend to be regional pollutants, not simply local pollutants.

Indoor Air Pollutants

Indoor air pollution is created partially because of inadequate ventilation. Pollution indoors may be exacerbated by high temperatures and humidity. The leading indoor air pollutant is environmental tobacco smoke (ETS). In developing countries that use wood smoke to heat homes and cook food, indoor air pollution is high. The pollutants include carbon monoxide, particulates, and other toxins. The World Health Organization estimates that 2.5 billion people are negatively affected by this smoke in a year.

Radon

Radon is a colorless, tasteless, radioactive gas naturally produced by the radioactive decay of uranium in bedrock. Radon accumulates in buildings with basements and buildings with slab foundations. Energy-efficient homes, which lack leaks, tend to have a greater risk of radon exposure. Increasing ventilation will decrease radon dangers, as will sealing cracks around plumbing, sewage, and gas connections in foundations. Radon is the second leading cause of lung cancer and the leading cause of lung cancer in nonsmokers.

Formaldehyde

Formaldehyde is released from building materials such as plywood, pressboard, textiles, furniture stuffing, and carpets. The human health effects of formaldehyde include dizziness, rashes, breathing problems, headaches, and nausea. To prevent formaldehyde exposure, other materials can be used in a building. For example, tile may be used instead of carpeting. Building materials and furniture can also be purchased that do not contain formaldehyde. Improved ventilation will also decrease exposure to formaldehyde.

Environmental Tobacco Smoke

Cigarette smoking is the leading cause of lung cancer, and exposure to secondhand smoke is the third leading cause of lung cancer, behind radon exposure. Cigarette smoking is also blamed for deaths from heart attacks, strokes, and other diseases induced by inhalation of the chemicals in environmental tobacco smoke (ETS). Banning smoking inside has resulted in a decrease in deaths due to secondhand smoke.

Asbestos

Asbestos is a naturally occurring fiber used because of its fire-retardant nature. It was used in insulation, ceiling tiles, rooting, and brake lining in automobiles. Once it was realized that asbestos induced the chronic lung disease asbestosis, lung cancer, and mesothelioma (a rare, fatal cancer), the EPA began to regulate its use, and it was phased out of use by 1997. The first response to asbestos in buildings was to remove it, which is more dangerous than leaving it in place. Only damaged, torn asbestos materials are dangerous because the small fibers separate and can be easily inhaled. Current asbestos policies are to leave the asbestos in place if it is intact or to seal it behind another material. If it has to be removed, only qualified individuals should be allowed to remove the asbestos.

Sick Building Syndrome

A sick building is a building in which deleterious health effects are linked to being in the building. These health effects include headache, eye and throat irritation, cough, dizziness, nausea, and fatigue. These symptoms improve once an individual leaves the building. A building is defined as sick if 20 percent of the occupants experience symptoms in the building but recover once they leave the building. Sick buildings are linked to inadequate ventilation, new buildings that contain chemicals such as formaldehyde from building products, or biological contaminants like mold or pollen.

Ozone Depletion

The ozone layer is present in the stratosphere. It is created when lightning causes oxygen gas to break into oxygen radicals (O^-). These radicals are unstable and bind with oxygen gas (O_2) to form ozone (O_3). The thinning of the ozone layer was first observed over Antarctica in September and October of 1985. In 1974 Mario Molina and Sherwood Rowland had demonstrated that chlorofluorocarbons (CFCs) had the ability to rise into the stratosphere where sunlight causes them to break down, releasing the Cl^-, which reacts with ozone, converting it into oxygen. They, along with Paul Crutzen, were awarded the Nobel Prize for Chemistry in 1995 for their research.

CFCs were widely used as refrigerants (Freon), coolants, and propellants. Halons is used in fire extinguishers, and carbon tetrachloride and methyl chloroform are used in a variety of industrial processes. Methyl bromide is used as a crop fumigant. The halons and methyl bromide release Br^-, even more damaging to ozone than Cl^-. Once it was realized the impact that ozone depletion was having, an international meeting occurred in Montreal in 1987. The meeting is now called the Montreal Protocol. It was agreed that nations cease the use of CFCs by the year 2000, although they later moved the date up to 1996 because viable alternative chemicals were available. The primary substitutes used are hydrochloroflourocarbons, which release less chlorine per molecule.

Human health effects of ozone depletion include increased cataracts, weakened immune systems, and greater incidence of skin cancers. Sensitive aquatic species also suffer from the increased UV radiation that now strikes earth. Intense UV radiation also damages plant tissues, resulting in reduced crop productivity and biomass in natural systems.

Acid Deposition

Acid deposition results from the deposition of the secondary air pollutants sulfuric, nitric, and carbonic acids. Sulfuric acid is responsible for about two-thirds of the acid deposition damage in the world and nitric acid the remaining one-third. In urban areas where NO_x are released by the numerous vehicles, the damage from each type of acid is nearly equal. The deposition may be rain, snow, fog, or dry deposition. Normal unpolluted rain has a pH of about 5.6 due to the natural levels of CO_2 in the air that create carbonic acid. Most acid deposition in the United States has a pH of about 4.3. Ecosystems based upon carbonate rocks have the ability to buffer the impact of the acids. The pH does not dramatically change in these ecosystems. Ecosystems present on igneous rock lack buffering ability and are far more susceptible to ecosystem acidification.

Vegetation is impacted by acid deposition. Acid directly damages cuticles (the waxy coat on leaves) on plant leaves, which makes the plants more susceptible to infections from bacteria, nematodes, and fungi. As a result, not only are forests damaged, but crop yields are reduced in acidified ecosystems. Forest damage is exemplified by the devastation in the Great Smokey Mountains in the United States and the Black Forest in Europe. Acid increases aluminum solubility and thus its toxicity, while leaching the important plant nutrients magnesium and calcium.

Acid deposition lowers the pH in aquatic systems and results in stress in sensitive organisms. The eggs and juveniles of sensitive species are susceptible to a pH of 5, thus interfering with reproduction. The acids therefore decrease biodiversity by reducing food available at lower trophic levels, which may have a resultant reduction of biomass. Sensitive fish species include game fish such as trout and salmon. Trash fish species, such as gar and carp, are more pollution tolerant.

Acid deposition also has an aesthetic impact on human structures. Buildings and statues made of carbonate rock sensitive to chemical weathering, such as limestone and marble, are etched and deeply damaged by years of acid deposition. Steel is corroded by exposure to these acids, as are paints and rubber.

Prevention of Air Pollution

The easiest way to prevent air pollution is to reduce the need for fossil fuels to be burned. Improving the efficiency of automobiles uses less gasoline. Efficiency can also be increased in home appliances, such as air conditioners, washing machines, dryers, and electrical heaters. Homes may also have more insulation and employ solar heating or cooling methods to decrease fossil fuel combustion. If a combustion energy source is required, natural gas is preferred over coal or oil because it releases less air pollution per unit of energy. Alternative energies, such as hydroelectric, geothermal, solar, and wind power may be used to alleviate air pollution. These mechanisms reduce all air pollutants.

Many pollution control strategies are dependent upon the pollutant. For example, sulfur dioxide may be reduced precombustion, during combustion, and postcombustion. Precombustion methods include using a higher grade of coal (anthracite) and washing the coal to remove excess sulfur. Coal may also be converted into a gas or oil, which removes the sulfur. A combustion method of removing sulfur from coal emissions is using fluidized bed combustion. Fluidized bed combustion is carried out by burning the crushed coal with crushed limestone. The sulfur in the coal combines with the calcium in the limestone to form calcium sulfate, or gypsum. This bottom ash can then be disposed of in a proper fashion. Postcombustion methods include using catalytic converters to oxidize the sulfur to yield sulfur compounds. A lime scrubber in a smokestack may also be used. In a wet scrubber, a slurry of lime mixed with water is sprayed across the exiting gases. The sulfur mixes with the calcium, forming the calcium sulfate, which falls to the bottom of the smokestack as bottom ash.

Particulates can be removed from coal combustion by burning coal that has been washed to decrease the ash content. Most particulate removal is postcombustion. The resultant particle mixture must often be discarded in a hazardous waste landfill. Bag filters are a series of bags, somewhat like a bag in vacuum cleaner, which catch the particulates as they rise in the smoke. The bags are periodically emptied of their ash. Electrostatic precipitators remove 99 percent of the particulates in coal emissions. They function by passing the coal emissions past a series of charged plates, thus charging the particulates, which then bind to an oppositely charged plate. Cyclone collectors create a vortex in a smokestack, causing the particles to collide and fall to the bottom of the stack as bottom ash.

Smog forming pollutants can be controlled in a variety of ways. NO_x can be decreased in any combustion system by controlling the combustion temperature. In automobiles, catalytic converters are used to promote complete combustion, which decreases NO_x, VOCs, and carbon monoxide. Many newer cars have systems that retrieve VOCs from the automobile engine and return them to the engine combustion chamber. This system is called a positive crankcase ventilation (PCV) system. One problem with reducing NO_x is that when you minimize their emissions, you tend to have an increase in VOCs. Therefore a balance is usually struck to minimize both as much as possible.

Legislation

The main legislation that controls air pollution is the Clean Air Act of 1963. It was dramatically amended in 1970 and 1990. The original act was designed to assist states in decreasing air pollution. The amendments created the list of criteria air pollutants and set the NAAQS. As needed, the NAAQS are modified by the EPA. For example, regional haze, caused by particulates, is being examined by the EPA. The cap and trade program devised by President Bush is designed to allow utilities not in compliance with EPA NAAQS to buy, sell, or trade pollution credits for sulfur dioxide, nitrogen dioxide, and particulates. The expense of these credits is a strong incentive to utilities to decrease their emissions to reduce their overall costs.

Developing Countries

As developing countries undergo demographic transition, they use more fossil fuels. These emissions are usually not subject to control because the goal is industrialization. Mexico City is one of the most polluted cities in the world. China has no air pollution controls on its coal burning plants, and thus the particulate, sulfur dioxide, and carbon dioxide emissions are high.

Chapter 16 Questions

Use the following for questions 1-4.

      a. mercury
      b. ozone
      c. formaldehyde
      d. carbon dioxide
      e. radon

1. may contribute to sick building syndrome
2. biomagnifies in trophic levels
3. is a secondary air pollutant
4. contributes to global warming

Use the following for questions 5-8.

      a. CFCs
      b. VOCs
      c. lead
      d. methane
      e. nitrogen dioxide

5. contribute(s) to the breakdown of ozone in the stratosphere
6. exposure in young children causes attention deficit, learning, and hyperactivity disorders
7. the majority of this air pollutant arises from natural sources
8. contributes to acid deposition

9. All of the following are criteria air pollutants except
a. lead.   b. mercury.   c. ozone.   d. carbon monoxide.   e. nitrogen dioxide.

10. The primary source of carbon dioxide in the atmosphere is
a. deforestation.   b. photosynthesis.   c. cellular respiration.
d. fossil fuel combustion.   e. biomass combustion.

11. Radon exposure causes
a. nausea.   b. breathing disorders.   c. lung cancer.
d. endocrine disruption.   e. attention deficit disorders.

12. Which of the following is correct regarding asbestos?
a. was used as a refrigerant and propellant
b. causes endocrine disorders when humans are exposed to it
c. removal is relatively safe and easy to accomplish
d. causes lung cancers and mesothelioma
e. causes kidney diseases in susceptible people

13. All of the following statements regarding stratospheric ozone are true except
a. ozone rises easily from the troposphere to the stratosphere.
b. ozone is formed when an oxygen radical created by the splitting of oxygen by lightning binds to oxygen gas.
c. chlorine radicals from chlorofluorocarbons cause ozone to break down into oxygen gas.
d. stratospheric ozone absorbs UV radiation.
e. the hole in the stratospheric ozone is easily observed in Antarctica.

14. Which of the following is not a greenhouse gas?
a. N_2O b. CO_2 c. CH_4 d. SO_2 e. H_2O vapor

15. Which of the following rocks are susceptible to damage due to acid deposition?
a. sandstone   b. granite   c. basalt   d. limestone   e. slate

16. Sulfur dioxide emissions can be prevented by all of the following except
a. lime scrubbers in a smokestack.
b. switching to bituminous coal or lignite.
c. converting coal into a gas or liquid.
d. fluidized bed combustion.
e. washing the coal prior to use.

17. The automobile component used to lower NO_x emissions is the
a. afterburner.   b. bag filter.   c. catalytic converter.
d. electrostatic precipitator.   e. scrubber.

18. Carbon monoxide
a. is one of the major greenhouse gases.
b. is released from incomplete combustion of biomass or fossil fuels.
c. causes lung cancer.
d. directly damages plant tissues.
e. causes increased crop yields.

19. All of the following lower particulate emissions except
a. lime scrubber.   b. bag filters.   c. electrostatic precipitators.
d. cyclone collectors.   e. fluidized bed combustion.

20. Mercury
a. bioaccumulates in bone.
b. is released into the atmosphere by gasoline combustion.
c. is converted into its most toxic form by bacteria.
d. causes lung cancer in individuals exposed to the metal.
e. is directly harmful to plants

Chapter 16 Answers

1. c. Formaldehyde is the pollutant that may contribute to sick building syndrome.
2. a. Mercury is the pollutant that biomagnifies in trophic levels.
3. b. The only pollutant listed that is a secondary air pollutant is ozone.
4. d. The only pollutant listed that contributes to global warming is carbon dioxide.
5. a. CFCs contribute to the breakdown of ozone in the stratosphere.
6. c. Lead exposure in young children causes attention deficit, learning, and hyperactivity disorders.
7. d. The majority of methane pollution arises from natural sources.
8. e. Nitrogen dioxide contributes to acid deposition.
9. b. Mercury is not a criteria air pollutant.
10. c. The primary source of carbon dioxide in the atmosphere is cellular respiration. Deforestation and combustion contribute to CO_2, but the natural production is far greater (90 percent) than the anthropogenic form. Photosynthesis would lower carbon dioxide in the atmosphere.
11. c. Radon exposure causes lung cancer due to the radiation exposure.
12. d. Asbestos causes lung cancers and mesothelioma. It was used as an insulator due to its flame-retardant nature. Its removal is dangerous because it is most dangerous when broken.
13. a. Ozone does not rise from the troposphere to the stratosphere.
14. d. Sulfur dioxide induces global cooling.
15. d. Limestone is susceptible to damage due to acid deposition because it is a carbonate rock.
16. b. Sulfur dioxide emissions can be prevented by switching to low S coal, like anthracite. Bituminous coal and lignite are high S coals.
17. c. The automobile component used to lower NO_x emissions is the catalytic converter.
18. b. Carbon monoxide is released from incomplete combustion of biomass or fossil fuels.
19. e. Fluidized bed combustion reduces SO_2 emissions.
20. c. Mercury is converted into its most toxic form by bacteria. Mercury bioaccumulates in fat and is released into the atmosphere by coal combustion.

Chapter 17 – Water Use and Management

Key Terms

aquifer	infiltration	water stress
artesian well	rain shadow	water table
condensation	recharge zone	withdrawal
condensation nuclei	relative humidity	zone of aeration
consumption	salt water intrusion	zone of saturation
desalination	saturation point	
dew point	sinkholes	
discharge	sublimation	
evaporation	subsidence	
groundwater	transpiration	

Skills

1. Diagram the water cycle.
2. Characterize groundwater and surface water features.
3. Appraise human water use.
4. Evaluate the priorities of water use for ecological communities, agricultural use, or municipal use.
5. Review water conservation methods.
6. Debate the costs and benefits of water diversion projects. Address the ecological and economic aspects of the projects.

Take Note: Understanding water use is important for the AP student. Numerous nations remain in conflict over water rights, and comprehension of basic water issues is imperative. Prior AP essay questions have asked students to explain the pros and cons of water diversion projects. The question expected students to be familiar with several different projects to answer the question. Water and its uses are often addressed in the multiple-choice questions.

Hydrologic Cycle

The hydrolic cycle is powered by the sun and gravity. Water is released from surface water through evaporation. Evapotranspiration is the loss of water from the leaves of a plant as they exchange gases necessary during photosynthesis. Sublimation is the conversion of solid water (ice) into the gaseous form directly, without a liquid stage. Once in the atmosphere the water molecules undergo condensation, and then precipitation occurs, returning the water to the earth. The water will infiltrate the soil and percolate down into the deeper layers of the soil or it will become runoff. The percolated water may become part of the groundwater, which flows steadily underground toward the ocean. The water may also be used by plants during photosynthesis. The runoff will become part of the surface water, entering lakes or flowing water systems. Relative humidity is the amount of water vapor in the air compared to the amount of water the air could hold at a given temperature (the saturation point). Water coalesces around particles called condensation nuclei to form clouds. The temperature at which water condenses is the dew point. Clouds release precipitation to reenter the cycle.

Rain is dependent upon atmospheric circulation, proximity to water, and topography. Areas with high levels of precipitation have atmospheric circulation that contributes to the rising and cooling of the warm moisture-laden surface air. These areas include regions with colliding air masses, the windward side of mountains, and around the equator. Sinking dry air occurs at 30° north and south of the equator and on the leeward side of mountains. In these regions, deserts form. The formation of deserts on the dry, leeward side of mountains is attributed to the rain shadow effect. The clouds release their moisture as they rise, so only the windward side of mountains receive precipitation, thus creating a "shadow" that lacks rain on the opposite side of the mountain.

Surface and Ground Water

Of all water on earth that is liquid water, 97 percent is in the ocean. Oceans moderate temperatures due to their current circulation and contain the majority of the earth's biomass. Most of the remaining fresh water is frozen in glaciers, icecaps, and snowfields. The largest of these ice sheets covers Antarctica.

After glaciers, groundwater makes up a large proportion of freshwater supplies. Groundwater is stored in porous underground rock, such as limestone or sand and gravel. This storage area is known as an aquifer. Aquifers may be surrounded by areas of impermeable rock (like shale or granite) or clay called aquicludes or aquitards. Such confined aquifers are typically under pressure and will readily flow when penetrated by drilling. These aquifers are called artesian aquifers. Other aquifers, known as unconfined aquifers, are not bounded at the top by an aquiclude and may form the water table. The water table is the highest level the water arises in the soil, thus forming the zone of saturation. The upper layer of soil is known as the zone of aeration, because the particles of soil are surrounded by air. Aquifers are filled in areas called recharge zones, which may not be near the aquifer. Groundwater does flow, and the rate of recharge must not exceed the rate of withdrawal or problems may occur. The largest aquifer in the United States is the Ogallala Aquifer, which lies primarily under Nebraska, Kansas, and north Texas. The water level in this aquifer has been dramatically lowered as water is removed for irrigation.

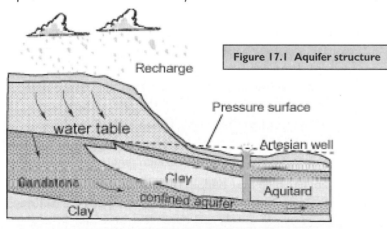

Recharge

Figure 17.1 Aquifer structure

Pressure surface

Artesian well

water table

Sandstone

Clay

confined aquifer

Aquitard

Clay

Water on the surface of the ground is called surface water, and it comprises less than 0.02 percent of all water. This water is in rivers, lakes, inland seas, streams, marshes, ponds, and swamps. The discharge of a river is the amount of water that passes a point in the water in a given amount of time. The river with the largest discharge is the Amazon, followed by the Orinoco. The North American river with the greatest discharge is the Mississippi River. Wetlands and swamps serve as recharge zones, purify water, and reduce erosion.

A watershed is the area of land that flows into a body of surface water. It is important that watersheds maintain their vegetation to reduce runoff and encourage infiltration.

Water Availability and Use

More than one-sixth of the world's population lack access to adequate clean drinking water and half of the world's population lacks access to sanitation. The World Health Organization considers 264,000 gallons of water per person to be minimum level below which shortages will impede development and damage human health. Two-thirds of the world's households do not have running water and must travel outside the home to wells to retrieve water.

Human intervention in the hydrologic cycle includes ground and surface water depletion, ground and surface water pollution, and the clearing of vegetation. The clearing of vegetation, particularly in temperate and tropical rainforests, interferes with the water cycle by decreasing transpiration.

Areas that have enough water include Brazil, Canada, Russia, and the Congo, because they have relatively large amounts of precipitation in conjunction with large amounts of land. High population densities and low precipitation result in water stress, or the lack of available water required to sustain a population. A country that uses more than 20 percent of its available water is experiencing water stress. Droughts can exacerbate water stress. Land use practices tend to exacerbate the effects of drought, as seen by the Dust Bowl in the United States in the 1930s. ENSOs contribute to droughts in many parts of the world.

Water withdrawal is removing water from surface water or groundwater for human use. Consumption is using the removed water in such a way that it is not useful again as surface water. Water that is withdrawn but not consumed may become degraded by chemical or thermal pollution. Water withdrawal is a concern, because as the human population increases, the demand for fresh water dramatically rises. The greatest consumer of water for human use is water used in irrigation and other agricultural needs. In the world, two-thirds of water withdrawal is used for irrigation.

In the United States the greatest volume of domestic water is used to flush toilets, followed by bathing, laundry, and dishes. Industrial processes use about 20 percent of the U.S. water withdrawals. Power production accounts for 50–70 percent of this use. The greatest producer of degraded water is mining.

Irrigation Methods

The original method of irrigation was to flood a field. Much of the water is lost to evaporation, so this method is inefficient. Sprinkler systems are also used, which also results in large amounts of water lost to evaporation. Low energy precision application (LEPA) methods use a center pivot sprinkler, which reduces evaporation because the sprinkler heads are lower to the ground. Drip irrigation involves using underground or surface pipes, which apply the water directly to the soil, further reducing water loss. These water conserving irrigation methods also prevent waterlogging and soil salinization.

Groundwater Withdrawal Problems

Groundwater is used for 40 percent of the municipal and agricultural needs in the United States. If water withdrawal exceeds recharge, several impacts occur. The water table can be dramatically lowered. If lakes are water table lakes, even surface water may be impacted by this loss of groundwater. Wells drilled into the aquifer may dry up, because the water table has been lowered. Sinkholes form when limestone dries out and crumbles due to excessive withdrawal. These sinkholes are localized damage. Sinkholes may also form when the water table rises again and limestone dissolves in the water. Sinkholes are prevalent in Florida around the Floridian Aquifer. They form quickly and have dropped cars and even houses into craters. Subsidence is the sinking of an area due to overdrawing of groundwater. The San Joaquin Valley in California is estimated to have sunk more than 30 feet in the last 50 years due to subsidence. Louisiana and Texas are subject to saltwater intrusion. This phenomenon occurs when fresh water is drawn out of an aquifer and salt water from coastal areas is pulled into the freshwater aquifer, thus contaminating it.

Water Diversion Projects

Water diversion projects move water from its original location to a new site. Dams alter ecosystems dramatically, by flooding upstream areas and decreasing flooding downstream. Dams impact the aesthetic value of a flowing water system and create a new standing water ecosystem. The benefits of dams are numerous. The major positive impacts are the regulation of water flow and flood prevention below the dam. The reservoir can provide irrigation and drinking water and hydroelectricity. The lake created by the dam can provide the recreational benefits of fishing, boating, skiing, and bird watching.

The costs of dams are also to be considered. Dams are expensive to build but have relatively low operating costs. They destroy riparian ecosystems. They increase the temperature of the water below the dam and decrease its dissolved oxygen. If the river flows to the ocean, the estuary becomes more salty due to the lack of influx of fresh water. There is also a decreased flow of nutrients to estuaries and nutrient-rich sediment downstream, important for maintenance of aquatic species downstream. The sediment that used to flow down the river collects in front of the dam, resulting in the need to dredge the reservoir. Reservoirs lose a tremendous amount of water to evaporation due to the large surface area.

Additionally, dams interfere with the spawning and thus survival of anadromous fish. Anadromous fish species are fish that spawn in fresh water and spend their adult life in the ocean. As they need to spawn, they return to the river in which they were hatched. These fish include salmon, sturgeon, and some trout species. Dams on rivers interfere with spawning because the mature fish cannot move upstream to breed. To alleviate this problem, fish ladders have been installed to allow the adult fish to bypass the dams. Several species of trout and salmon are bred in hatcheries to increase their numbers. The smolt die going downstream as they pass through the hydroelectric turbines. To alleviate this problem, smolt and fry from hatcheries are trucked around dams and placed into the water below the dam. Some strains of Coho, Chinook, and Sockeye salmon and some steelhead trout have been placed on the endangered species list and are managed by a recovery plan by the NMFS because they are anadromous fishes.

Specific Water Diversion Projects

The Everglades is a 50-mile wide river that flows through south Florida toward Florida Bay. In the 1960s the U.S. Army Corps of Engineers wanted to help drain the Everglades for agriculture and development and to alleviate damage from flooding. To drain the area, the Kissimmee River, which drained the wetlands around Lake Okeechobee, was straightened via canals from its original meandering state. The state of Florida, in conjunction with the federal government, is cleaning up the mercury and phosphate pollution in the Everglades and restoring the natural meanders to the Kissimmee River.

The fourth largest inland sea has become a salty desert. The Aral Sea, bounded by Kazakhstan and Uzbekistan, has shrunk more than 60 percent of its water volume and 75 percent of its size since the 1960s. The sea was fed by the Amu Darya and Syr Darya rivers, diverted for irrigation of cotton and rice. As the waters feeding the sea were diverted and evaporation continued, the sea became more and more salty, destroying lucrative fishery. Corroded ships sit in sand, miles from water. The health issues in the region include lung diseases from salt storms, anemia, and thyroid and kidney diseases. The infant mortality and cancer rates have increased dramatically. Mono Lake in California has suffered from water diversions, as well. This lake has lost one-third of its surface area, destroying the habitat of migratory and wading birds.

The James Bay project built in Canada diverted three rivers from the Hudson Bay and created lakes in forests and tundra. Caribou drowned trying to follow migratory paths across flooded regions, and indigenous people, the Cree, were harmed due to loss of traditional hunting and fishing grounds. The soil was contaminated with mercury in the area, and the flood waters caused the mercury to enter the food chain.

When the flow of the Nile River was impeded by the building of the High Dam in Aswan, Egypt, in the 1960s, schistosomiasis cases increased in the area. The dam created Lake Nassar, a body of standing water in which the snails required to transmit schistosoma parasites could thrive. The snails cannot survive in flowing water. A similar situation arose on the Senegal River due to the Diama Dam and could perhaps be a problem at Three Gorges. The dam is being built in an area endemic for schistosomiasis, so increasing the standing water is likely to increase the disease incidence.

The Three Gorges Dam, under construction across the Yangtze River in China, will be the world's largest hydroelectric dam. This dam is covering ancient cities and is displacing over 1 million people. The dam is built along a fault, and grave concerns regarding its safety have not been adequately addressed.

Desalination

Desalination is the process of removing water from salty or brackish water, creating fresh water and a salty brine. One process used to carry out desalination is reverse osmosis, in which energy is used to force water through a semipermeable membrane against the concentration gradient. Distillation may also be used, in which the water is heated to boil out the salt, then condensed as fresh water. This method has been used a lot in the Middle East due to the lack of water in the region. Water produced by desalination is about four times more expensive than other water. In the United States, many regions are turning to desalination due to water shortage. For example, a desalination plant was built in Tampa, Florida, to accommodate the burgeoning population and its demand for fresh water. The plant uses the waste heat from an electrical plant (cogeneration) to power reverse osmosis. Perhaps the greatest environmental impact of desalination is the disposal of the salty brine.

Other Ways to Increase Water Supplies

Many other ways have been suggested to increase freshwater supplies. Cloud seeding with salts or dry ice has been recommended to increase precipitation. However, the hydrologic cycle in some areas makes this idea extremely difficult to implement. Harvesting icebergs from the poles has been recommended, but the impacts to warmer regions will need to be taken into consideration if such a method is employed. Another method currently being examined is called aquifer storage and recovery. In this method, surface runoff is collected, treated, and pumped into aquifers for future use.

Water Conservation

The easiest method to conserve water is to use less. Flushing toilets is the primary municipal water use in the United States, so low-flow toilets may be used, which use 1.6 gallons per flush as opposed to the older models, which used up to 8 gallons of water. Low-flow shower heads and high-efficiency washers and dishwashers dramatically reduce water use. To encourage consumer water conservation, they should be given rebates for purchasing water conserving appliances, toilets, and shower heads. Municipalities can also charge consumers more for their water to reflect water shortages. Water subsidies can be decreased and watershed management can be better controlled to alleviate water problems. Consumer education regarding water conservation strategies is imperative to encourage compliance. Industrial water can be recycled to decrease water use. Agricultural water waste can be decreased by more efficient irrigation methods.

One method to conserve municipal water used in irrigation is to have nonpotable reuse of the water. Gray water is water collected in a house from showers and sinks. This water does not enter the sewer or septic tank, but is collected to water lawns and gardens. A similar reuse is reclaimed water, in which the city purifies sewage effluent to the point of discharge and provides it as irrigation water. The positive impacts of reclaimed water are numerous. There is no need to use potable water for irrigation, nor does the sewage effluent have to be dumped into surface water. Additionally the water can naturally percolate into soil, increasing groundwater stores. Arid areas can also employ xeriscaping, or planting native species that require little to no watering.

Chapter 17 Questions

1. Stream discharge is
a. the length of a river times its deepest point.
b. the distance the water flows from the headwater to the mouth.
c. the speed the river travels past a fixed point.
d. increased in the summer due to increased temperatures.
e. the volume of water the river holds at any given time.

2. Materials that prevent the flow of groundwater are called
a. aquifers.   b. aquicludes.   c. wells.   d. limestone.   e. artesian.

3. Removal of excessive amounts of groundwater in coastal areas may result in
a. saltwater intrusion.   b. permeablility damage.   c. subsidence.
d. aquifer recharge.   e. surface water depletion.

4. All of the following processes involve water moving with gravity except
a. infiltration.   b. percolation.   c. precipitation.   d. transpiration.   e. runoff.

5. Which of the following water diversion is correctly matched with its problem?
a. increased schistosomiasis in the Aral Sea region
b. increased saltiness of the Yangtze in China due to the Three Gorges Dam
c. reduced water flow through the Everglades has caused a reduction of biodiversity
d. Mono Lake in California experiences flooding due to irrigation canal placement
e. James Bay has increased in saltiness due to the water diversion for irrigation purposes

6. The greatest amount of fresh water is found in
a. groundwater.   b. inland seas.   c. lakes and ponds.
d. ice and snow.   e. rivers and streams.

7. Which of the following is not an impact of dams?
a. greater flooding below the dam
b. impeded breeding in anadromous fishes
c. reduced sediment flow downstream
d. lowered dissolved oxygen in the water downstream
e. increased salinity in estuaries fed by dammed rivers

8. Which of the following policies would be a disincentive, or "stick," that would encourage water conservation in a municipality?
a. increasing water costs to reflect water shortages
b. increasing the availability of reclaimed water
c. providing rebates for low-flow toilets and shower heads
d. providing rain barrels to catch rainwater for watering lawns
e. providing a property tax break on homes that use xeriscaping

9. Sinkholes result from
a. saltwater intrusion.   b. raising the water table.
c. increased evaporation of groundwater.   d. excessive removal of groundwater.
e. increased flow of surface water during the spring after snowmelt occurs.

10. The greatest use in municipalities of fresh water is
a. bathing.   b. washing dishes.   c. flushing toilets.
d. laundry.   e. food preparation.

Chapter 17 Answers

1. c. Stream discharge is the speed the river travels past a fixed point.
2. b. Materials that prevent the flow of groundwater are called aquicludes.
3. a. Removal of excessive amounts of groundwater in coastal areas may result in saltwater intrusion.
4. d. Transpiration is the movement of water vapor from a tree's leaves into the atmosphere.
5. c. Reduced water flow through the Everglades has caused a reduction of biodiversity.
6. d. The greatest amount of fresh water is found in ice and snow.
7. a. There is less flooding below a dam.
8. a. Increasing water costs to reflect water shortages would be a disincentive, or "stick," that would encourage water conservation in a municipality. The other choices would be incentives, or "carrots."
9. d. Sinkholes result from excessive removal of groundwater.
10. c. The greatest use in municipalities of fresh water is flushing toilets.

Chapter 18 - Water Pollution

Key Terms

atmospheric deposition	fecal coliforms	red tide
biological oxygen demand	nonpoint source	secondary water treatment
cultural eutrophication	oligotrophic	tertiary water treatment
dissolved oxygen	oxygen sag curve	total maximum daily loads
effluent sewerage	point source	
eutrophic	primary water treatment	

Skills

1. Characterize the different types of water pollutants, including their source, environmental impact, and human health effects.
2. Relate sewage treatment and drinking water quality to health, safety, and quality of life.
3. Summarize the quality of water in developed and developing countries.
4. Recall contamination problems associated with groundwater.
5. Identify and describe the major ocean pollutants.
6. Appraise the effectiveness of the Clean Water Act and the Safe Drinking Water Act.

Take Note: You must be familiar with the common water pollutants, their origin and impacts on the environment and human health, as well as ways to remediate the pollutant. For example, one essay question expected students to understand how mercury entered aquatic ecosystems, where the mercury originated, how the mercury could be prevented from entering the ecosystem, and the impacts of bioaccumulation of mercury.

Water Pollution

For generations, Americans dumped toxic chemicals, wastes, solvents, and sewage into rivers and lakes. To protect our waterways, President Nixon signed the Clean Water Act in 1972. The act has been modified several times, but the main goal is to protect surface and groundwater from pollutants. Pollutants are any physical, chemical, or biological changes in water quality that adversely affects living organisms.

Point source pollution is any pollution derived from an easily identifiable source, such as a drain pipe, smoke stack, or factory. Nonpoint source pollution lacks an easily identifiable source. Examples include agricultural runoff, construction site runoff, urban storm runoff, and feedlot runoff. Atmospheric deposition of chemicals may also be considered nonpoint source pollution.

Flowing water systems resist contamination by pollutants because the flowing water tends to dilute and distribute the pollutant. The rate at which the system can clean itself is related to the volume of the river, the rate at which the river flows, and the temperature of the river. A standing water system is far more susceptible to damage because there is no way for the pollutant to flow away.

Take Note: It is imperative that you are familiar with the tests conducted on water to determine water quality. For example, past AP essays have asked students to identify and describe water tests that explain the water quality of a particular body of water. You must be able to differentiate between physical, chemical, and biological water tests and know if those tests are biotic or abiotic.

Water Quality Tests

Numerous tests may be conducted upon water to determine the quality. The turbidity, or cloudiness, of the water will be an indicator of the ability of the algae and emergent vegetation to carry out photosynthesis. The turbidity is increased due to sediment pollution. The color of the water may also play a role. Water that contains a lot of tannins from leaves cause the water to be very dark, and such water is called "black water." Dark water may impair photosynthetic activity. The pH level of water can be taken to determine if the water is too acidic or basic. Extreme pH levels will kill sensitive species. Salinity plays a role in water quality. The levels of magnesium and calcium ions, or the measure of hardness in fresh water, can also be determined. The levels of the inorganic nutrients nitrates and phosphates can be measured to determine water quality. The level of dissolved oxygen (DO) can be measured. The DO levels will determine the animal species capable of surviving in an aquatic ecosystem. The biological oxygen demand, BOD, can be measured as well. The BOD is the amount of oxygen required to sustain an aquatic system. The BOD will rise in the presence of decomposing organic matter (known as oxygen demanding wastes), resulting in a decrease in the DO. Several other factors affect levels of dissolved oxygen. For example, temperature is influential because cold water retains more oxygen than warmer water. Light penetration, turbidity, and color also affect DO because with more light there is more photosynthesis, hence more available oxygen. Turbulence increases the DO levels due to the mixing of atmospheric oxygen into the water. The presence of moderate amounts of submerged vegetation, emergent vegetation, and algae increases the DO.

Analysis of the levels of benthic macroinvertebrates may also be done to determine water quality. In studies of stream benthic macroinvertebrates, mayfly, dobsonfly, and caddisfly larvae are indicative of excellent water quality whereas the presence of blood worms, leeches, and pouch snails are indicative of poor water quality. Many species of fish are sensitive to pollutants or poor water quality and will not be present in contaminated areas. The biodiversity of an area can be assessed to determine if contamination is occurring. Fecal coliform counts may also be made on water sources to determine if fecal contamination is occurring. Fecal coliforms are bacteria such as *Escherichia coli* (*E. coli*), an inhabitant of human intestines. The coliforms are not necessarily dangerous, but their presence indicates contamination with human or animal wastes that may contain pathogenic organisms. Drinking water may not contain any coliforms, and swimming water may not contain more than 200 coliforms in 100 ml of water tested. Beaches, lakes, rivers, and swimming pools are closed if the fecal coliforms exceed safe swimming levels and only reopen when the levels drop to an acceptable amount.

An oxygen sag curve illustrates the changes that occur in a flowing system when oxygen demanding wastes are added. The area upstream of the wastes exhibits good water quality, but when the wastes are added, the DO drops due to the increased BOD, resulting in a dramatic shift in biodiversity. As the water flows, the system slowly recovers back to its original state.

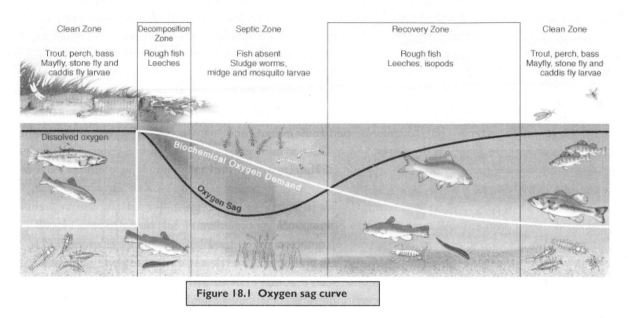

Figure 18.1 Oxygen sag curve

Types of Water Pollutants

Sediment pollution is the greatest pollutant by volume in aquatic systems. The sediment arises primarily from erosion of the land surrounding the water. This erosion may arise from mining, deforestation, rangeland, construction, and agriculture. Sediment may cover gravel where fish lay their eggs and may clog the gills of filter feeders. Sediment increases the turbidity, which impairs photosynthesis.

Pathogenic organisms are the water pollutants that pose the greatest threat to human health. Waterborne diseases include typhoid, cholera, hepatitis A, polio, schistosomiasis, and a variety of dysentery diseases. Mosquitoes that breed in water transmit dengue, malaria, yellow fever, and a variety of encephalitis viruses. Human wastes are the main source of these pathogens, but animal wastes from feedlots or food processing plants may also increase these bacteria in water supplies. In developed countries sewage treatment is readily available, and drinking water is treated to kill pathogens prior to its dispersal to the public. It is estimated that 2.5 billion people lack adequate sanitation in developing countries and over 1 billion lack safe drinking water.

Water can also be contaminated by excessive amounts of nutrients. A body of water low in nutrients is called oligotrophic, and a body of water high in nutrients is called eutrophic. The process of a body of water proceeding from nutrient poor state to a eutrophic state is called eutrophication and is a natural occurrence. When humans exacerbate the rate at which the water becomes eutrophic, it is called cultural eutrophication. Cultural eutrophication is typically caused by excess nutrients from fertilizer runoff or sewage. When excess nutrients are present, they induce an algal bloom. These algae cover the water so thickly that sunlight cannot penetrate to the bottom vegetation and lower algae, which kills them. Their decay, which increases the BOD, consumes oxygen required for aerobic heterotrophs to survive due to the hypoxic state of the water. These animals die, requiring further decay, thus consuming more oxygen in a positive feedback cycle. The sewage contamination also consumes oxygen due to the higher BOD. Another type of algal bloom associated with runoff is red tide. Red tide is a proliferation of certain species of dinoflagellates. These algae release a toxin lethal to numerous species of fish and has been attributed to killing West Indian Manatees in Florida. The toxin is an irritant to human eyes and respiratory systems. A noted red tide agent, *Pfiesteria piscicida*, even attaches directly to fish and kills them.

Inorganic pollutants are common in aquatic systems. These pollutants include metals, acids, bases, and salts. The heavy metals, including mercury, lead, tin, and cadmium are of great concern as are the super toxic metals selenium and arsenic. The heavy metals are persistent and tend to bioaccumulate and biomagnify. These metals typically arise from mining, mine waste, and smelting. Mercury is added to the atmosphere by coal combustion, incineration, and smelting. Cadmium increases due to mining and smelting. Tin was used on ships as an antifouling agent until its use was banned due to concerns of toxicity. Lead is the heavy metal contaminant that has declined dramatically in the United States since the banning of leaded gasoline, leaded paint, and lead pipes and solder. The maximum legal level of lead in drinking water is 20 ppb.

The nonmetallic salts are dangerous because they are soluble in water. Salts that contain selenium and arsenic build up to dangerous levels in desert ecosystems. Irrigation has increased the salt levels in some areas. In the northern United States, road salt is used to melt ice on bridges and roads. This corrosive salt makes its way into ecosystems.

Acids arise from smelting, metal plating, petroleum distillation, and coal mining (acid mine drainage). Acids also form in the atmosphere from the gases released during fossil fuel combustion. Older soils or soils derived from igneous rock cannot buffer the acid in ecosystems, but limestone and other carbonate rocks have the ability to decrease the impact of acids in aquatic systems. If the pH drops too low, biodiversity is limited to acid-resistant mosses. The acidic conditions also make aluminum more soluble, thus increasing its toxicity.

Synthetic organic chemicals also contaminate aquatic systems. Some of the examples of these chemicals include DDT, dioxins, PCBs, and other chlorinated hydrocarbon chemicals. The two main sources of these chemicals are improper waste disposal and agricultural and urban pesticide runoff.

Thermal pollution, or thermal shock, is due to the heating of water, which is then expelled back into surface water systems. This water is used to cool plants that use steam turbines to generate electricity or to cool other heat-producing industrial processes. The heated water has a negative impact on temperature sensitive species, usually organisms that occupy lower trophic levels. The hotter water also decreases the levels of DO in the water. Larger animals, such as game fish, birds, and manatees, thrive in the warm water effluent from such processes. Water may also increase in temperature when humans remove the existing vegetation from around a body of water. To prevent thermal shock, many power plants have cooling ponds or towers that cool the water prior to being released into natural systems.

Current Water Quality

The Clean Water Act was passed in 1972 and amended in 1977 to decrease pollution in U.S. waters. The act established discharge permits for waste dumping in surface waters. The purpose of the permits was to decrease water pollution from industry or municipal point sources, particularly from sewage treatment plants. The act also required best practicable control technology and set best available economically achievable technology goals for toxics and other pollutants.

The act allows the EPA to set wastewater standards for industry and set limits on pollutants. The act also has been used to protect wetlands, because it regulates the draining and filling of wetlands. In 1998, the EPA began to focus on watershed quality in order to protect surface water from contamination. States have to identify bodies of water that do not meet goals and develop total maximum daily loads (TMDLs) on each pollutant for each body of water. Future TMDLs will include pollution from acid deposition and background pollutants.

Developing countries and poorly developed countries do not spend money on sewage treatment, and many countries do not adequately provide safe drinking water. As a result, deaths from infectious disease remain high, particularly in the young.

Groundwater Quality

About half of the U.S. drinking water is derived from groundwater. In the past it was believed that percolation of water through soil as it moved into groundwater reserves purified the water. We know now that contaminants frequently remain in water and are therefore even more difficult to treat. Currently in the United States, groundwater may be contaminated by fertilizers and pesticides. Nitrates are extremely dangerous for infants to ingest, as it impairs their ability to carry oxygen. Aging underground storage tanks contaminate groundwater with a variety of chemicals including gasoline. The gasoline additive MTBE promotes complete combustion and therefore reduces CO and VOC emissions from automobile exhaust. Unfortunately MTBE is a suspected carcinogen that readily enters groundwater and does not degrade in the anoxic conditions in an aquifer. Leachate from landfills also enters groundwater if adequate precautions are not taken.

Ocean Pollution

Major contaminants in ocean waters are plastics and oil. Plastics kill organisms directly by wrapping around their bodies or by blocking their digestive systems when ingested. Plastics that are bio- or photodegradable are not as dangerous to marine life because they break down more rapidly than regular plastics. Oil originates naturally from seepage, but also is derived from bilge pumping, oil rigs, tanker accidents, and tank cleaning. Land is also a source of oil because urban runoff contains oil and gasoline. The 1990 London Dumping Convention proposed the Law of the Sea Treaty, which states that industrial waste, tank washing effluent, and plastic trash could not be dumped in the ocean after 1995. There are 64 countries, including the United States, that have agreed to the treaty.

Control of Water Pollution

The most inexpensive way to control water pollution is to prevent it. Many pollutants, such as lead in gasoline and DDT, have been banned due to their toxicity and therefore are no longer a threat to U.S. waterways. Nonpoint sources of pollution are difficult to control. Agriculture runoff contains pesticides, fertilizer, sediment, and animal wastes from feedlots. Urban runoff contains pesticides, fertilizer, sediment, oil, salts, heavy metals, and pet fecal matter. Construction sites result in sediment pollution. To prevent nonpoint pollution of waterways, best management practices have been developed. For example, animal wastes may be collected in a lagoon and the water passed through a wetland to remove bacteria and excess nutrients prior to discharge into open water. Applying only necessary levels of pesticides and fertilizers will result in decreasing amounts of these materials in runoff. Streets may be swept regularly to prevent trash and oils from entering urban runoff.

The Comprehensive Environmental Response, Compensation, and Liability Act (Superfund Act) and the subsequent Superfund Amendments and Reauthorization Act of 1984 provides funding for cleanup of contaminated abandoned sites that may be affecting water quality.

Sewage Treatment

In many countries with a low-population density, sewage treatment is unnecessary. In areas with high-population densities, treatment is necessary to prevent the transmission of disease. A lack of adequate sanitation in Mexico City, coupled with a burgeoning population, has resulted in fecal snow. Fecal snow is dried human fecal matter that becomes airborne particulate matter. Human feces are often used in developing countries as a source of fertilizer. These feces can carry infectious disease, and thus ingesting fruits and vegetables grown in these areas may be hazardous to one's health. In the U.S. outhouses were common until the 1950s, when septic tank systems and their drainage fields allowed for individual homes to treat wastes without risk of disease or contaminating groundwater.

Septic tanks function by receiving wastes from a household. Oils and grease float to the top and the solids fall to the bottom, where they undergo decomposition. The liquids flow through a series of pipes under the ground and are aerated to kill the usually anaerobic pathogens. The liquids are then taken up by the surrounding soils. The septic tanks must be pumped out occasionally to remove the solids.

Some areas are using effluent sewerage, which has a tank to digest solid waste but instead of using a drainfield for the effluent, the effluent is pumped to a treatment plant. Other areas release wastes directly into a constructed wetland to remove the wastes.

When population densities increase, sewage treatment plants are required to handle the large amount of human wastes. Storm sewers may combine with sanitary sewage forming combined sewer systems. There are two stages that occur in sewage treatment and a third is employed in many areas. The first stage is called primary treatment (Figure 18.2, a). The primary treatment removes the solid portion of the wastes in a mechanical process. The first step is to pass the sewage through grates to removed large debris like rags. A grit tank follows, which allows the large bits of sand and gravel to settle. These grit tanks are important in areas where a combined sewer system is present. The next step is to remove the organic solids separate from the liquid portion by allowing the solids to settle in a settling tank as sludge. Secondary treatment (b) involves removing the dissolved organic components of the wastes using bacteria. Several different methods exist for secondary treatment. The trickling filter bed passes the liquid portion of the waste over a layer of stones covered in bacteria, which degrade the organic material in the wastes. The aeration tank digestion, also called activated sludge process, mixes the effluent from the primary treatment with bacteria and oxygen to promote decomposition of the wastes. The effluent from the secondary treatment is then disinfected with chlorine, ozone, or UV light to remove bacteria and then released into a nearby waterway.

Tertiary water treatment (c) involves running the secondary treatment effluent through a wetland to remove the excess nutrients such as nitrates and phosphates.

In some areas, the effluent from the secondary treatment is processed a bit farther, then sent out, under pressure, to homes to be used for watering yards. This water is known as reclaimed water and solves several water issues. The reclaimed water does not require effluent entering surface water supplies. It also allows slow recharge of groundwater by allowing the water to trickle through the soil of the lawns on which it is being used. Potable water is not being used to water yards.

The sewage sludge was once dumped into the ocean, but now is incinerated, buried in a landfill, composted, or subjected to further anaerobic digestion to be used as a soil conditioner. Toxins and heavy metals in the sludge determine its final fate, because toxic materials are not suitable as compost or soil conditioners.

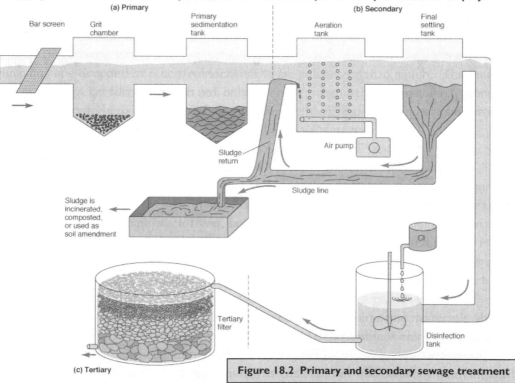

Figure 18.2 Primary and secondary sewage treatment

Potable water

Potable, or drinking, water is managed under the Safe Drinking Water Act, which regulates water quality in commercial and municipal systems. In accordance with the act, the EPA has set maximum contaminant levels for 90 different pollutants in drinking water. To generate potable water, the first step is to clarify the water by adding coagulants, chemicals that will coagulate dirt and other particles by causing them to stick together and sink. The water is then passed through a filter to remove many of the disease-causing agents. The water is disinfected, usually with chlorine, to ensure the safety of the water. Water may also be disinfected using UV light or ozone. Many areas add fluoride to increase dental heath. The maximum allowable level of fluoride in drinking water is 4 mg/L as established by the EPA.

Chapter 18 Questions

Use the following for questions 1-4.

      a. fecal coliforms
      b. nitrates
      c. atrazine
      d. sediment
      e. methyl mercury

1. released from combustion of coal; tends to bioaccumulate and biomagnify
2. indicator that human pathogens may be present in aquatic systems
3. broad-leaf herbicide frequently found in surface water and groundwater
4. pollutant that would induce eutrophication

Use the following for questions 5-8.

      a. Biological Oxygen Demand
      b. pH
      c. hardness
      d. turbidity
      e. phosphates

5. measurement increases in the presence of sediment pollution
6. determines the concentration of hydrogen ions in an aquatic system
7. would induce eutrophication in a standing water system
8. increases when decomposition increases

9. Which of the following are biological processes in the treatment of sewage?
      I. primary sewage treatment
      II. secondary sewage treatment
      III. tertiary sewage treatment

a. I only   b. II only   c. III only   d. I and II   e. II and III

10. All of the following are point sources of pollution except.
a. heated effluent from a coal power plant cooling system.
b. a steel factory emitting carbon dioxide.
c. rain washing pesticides from a field into a nearby lake.
d. automobile exhaust pipe emitting carbon monoxide.
e. sewage effluent pipe releasing treated wastewater into surface water.

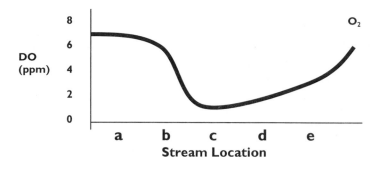

Questions 11-14 refer to the above diagram of an oxygen sag curve due to untreated human sewage.

11. Which location on the diagram indicates the highest biological oxygen demand?
12. Which location would you expect to find midgefly larvae, blackfly larvae, pouch snails, and leeches?
13. Which location would represent the introduction of untreated human sewage?
14. Which parameters would you expect to be increased at point c?

        I. temperature
        II. turbidity
        III. pH

a. I only   b. II only   c. III only   d. I and II   e. I, II, and III

15. Feedlots contribute which of the following to surface water?

        I. sediment pollution
        II. fecal coliforms
        III. nitrates

a. I only   b. II only   c. III only   d. I and II   e. I, II, and III

16. All of the following are persistent organic pollutants except
a. polychlorinated biphenyls.   b. DDT.   c. dioxins.
d. organochlorine pesticides.   e. nitrates.

17. Acidification in ecosystems can arise from all of the following except
a. surface water flowing through abandoned coal mines.
b. urban runoff containing high levels of salts entering surface water systems.
c. nitrogen oxides from automobile exhaust forming nitric acid in the atmosphere.
d. sulfur dioxide from coal burning power plants forming sulfuric acid in the atmosphere.
e. rain carrying sulfur from mine tailings into surface water.

18. Nitrate levels in water are addressed by which of the following legislative acts?
    I. Safe Drinking Water Act
    II. Clean Water Act
    III. Superfund Act
a. I only   b. II only   c. I and II   d. II and III   e. I, II, and III

19. Which of the following contaminants were found in groundwater once lead was banned from gasoline?
a. mercury   b. MTBE   c. DDT   d. nitrates   e. N₂O

20. All of the following are used in the United States to dispose of sewage sludge except
a. ocean dumping.   b. incineration.   c. treatment to become soil conditioner.
d. compost.   e. land fill burial.

Chapter 18 Answers
1. e. Methyl mercury is released from combustion of coal, and it bioaccumulates and biomagnifies.
2. a. Fecal coliforms indicate that human pathogens may be present in aquatic systems.
3. c. Atrazine is the broad-leaf herbicide frequently found in surface water and groundwater.
4. b. Nitrates, along with phosphates, are pollutants that would induce eutrophication.
5. d. Turbidity is a measurement of the cloudiness of water and therefore would increase in the presence of sediment pollution.
6. b. pH measures the concentration of hydrogen ions in an aquatic system. pH is the negative logarithm of the hydrogen ion concentration. For example, pH of 4 has an H⁺ concentration of 1 x 10⁻⁴.
7. e. Phosphates could induce eutrophication in a standing water system.
8. a. The Biological Oxygen Demand increases when decomposition increases due to the increased need for oxygen for decay.
9. e. Secondary and tertiary treatments are biological processes in the treatment of sewage. Primary treatment is a physical process.
10. c. Rain washing pesticides from a field into a nearby lake is nonpoint pollution.
11. c. The BOD is highest where the DO is the lowest because a high biological oxygen demand results in low DO.
12. c. Midgefly larvae, blackfly larvae, pouch snails, and leeches are indicative of poor water quality and would be expected at point c.
13. b. The DO begins to drop, then drops dramatically at point c, indicative of high levels of oxygen demanding wastes.
14. d. Temperature and turbidity tend to increase with high levels of oxygen demanding wastes.
15. e. Feedlots contribute sediments, coliforms, and nitrates to surface water.
16. e. Nitrates are inorganic.
17. b. Salts do not contribute to acidification.
18. c. The Superfund Act refers to cleanup of hazardous waste sites, not nitrate pollution.
19. b. MTBE began to be used as an oxygenate in fuel once lead was banned and began appearing in water supplies.
20. a. Ocean dumping is illegal in the United States.

Chapter 19 – Conventional Energy

Key Terms

black lung disease	fuel assembly	power
breeder reactor	joule	proven resource
chain reaction	methane hydrate	tar sand
control rod	nuclear fission	work
energy	nuclear fusion	Yucca Mountain
fossil fuel	oil shale	

Skills

1. Characterize current energy resources used in the United States.
2. Contrast U.S. use of energy versus energy use in other nations.
3. Classify the reserves of fossil fuels found in different nations.
4. Evaluate the costs, both economic and environmental, of using fossil fuels.
5. Evaluate the benefits, both economic and environmental, of using fossil fuels.
6. Appraise the costs and benefits of nuclear energy.
7. Diagram a light water reactor and know the functions of all components, including the water systems.
8. Evaluate proposed methods for nuclear waste disposal.

> **Take Note:** Nearly every AP exam contains an essay question about energy. These questions have included solar energy, coal, natural gas, biomass, and nuclear power, to name a few. It is expected that a student can carry out simple algebraic calculations on energy conversions, costs, and emissions values. You must be familiar with the recovery of the nonrenewable resource, the environmental damage from the recovery, the use of the resource, and the environmental damage that results from the use of the resource. Be sure that you know specifically which pollutant arises from which fossil fuel. You must also know ways to decrease the pollutants that arise from use of these fossil fuels. Energy questions will also be prevalent in the multiple-choice portion of the AP exam.

Energy

Energy is defined as the ability to do work. Work is force acting across a distance. Power is the rate at which work is done. Food energy is measured in calories. A calorie is the amount of energy that can raise the temperature of 1 gram of water 1°C. A kilocalorie, denoted calorie by most individuals, is 1000 calories. A newton is the force needed to accelerate 1 kg of mass 1 meter/second.

> **Take Note:** See table 19.1 for a list of energy units you should be familiar with using in calculations.

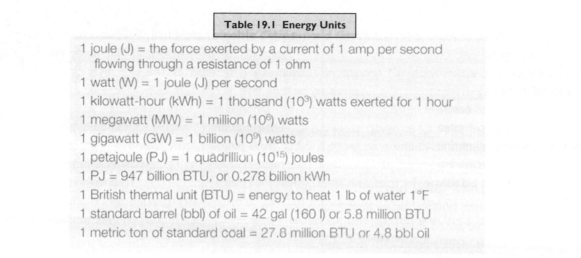

Table 19.1 Energy Units

1 joule (J) = the force exerted by a current of 1 amp per second flowing through a resistance of 1 ohm

1 watt (W) = 1 joule (J) per second

1 kilowatt-hour (kWh) = 1 thousand (10^3) watts exerted for 1 hour

1 megawatt (MW) = 1 million (10^6) watts

1 gigawatt (GW) = 1 billion (10^9) watts

1 petajoule (PJ) = 1 quadrillion (10^{15}) joules

1 PJ = 947 billion BTU, or 0.278 billion kWh

1 British thermal unit (BTU) = energy to heat 1 lb of water 1°F

1 standard barrel (bbl) of oil = 42 gal (160 l) or 5.8 million BTU

1 metric ton of standard coal = 27.8 million BTU or 4.8 bbl oil

Humans began using domesticated animals more than 10,000 years ago to assist us with our work. Wind and water were used to grind grain, cut timber, and provide other necessary energy. Steam engine development reduced available wood supplies and increased the use of coal. Coal fell into decline when it was discovered that petroleum could be used for many of the same applications. Eighty-six percent of the world's energy use is supplied by fossil fuels.

In the past, developed countries have used far more energy than developing countries. With the rapid industrialization and economic growth in China and India, developing countries will consume a greater proportion of energy than they have in the past.

Political turmoil in the Middle East has impacted fuel prices in the past. The effects were far more dramatic in developing than developed countries. The two events in the 1970s that affected prices were the oil embargo by OPEC (Organization of Petroleum Exporting Countries) in 1973 and the Iranian Revolution in 1979. The mid-2000s increase in gasoline prices were linked to increased worldwide demand for petroleum; increased price of petroleum; and damage done to refineries, pipelines, and U.S. rigs in the Gulf of Mexico due to Hurricane Katrina.

Figure 19.1 World energy use

Renewable 7%
Nuclear 7%
Oil 40%
Natural gas 23%
Coal 23%

American Energy Use

The United States consumes oil for 43 percent of our energy use. Due to the vast coal reserves in the United States, we use more coal than natural gas, even though coal combustion is far more polluting. Nuclear power accounts for 8 percent of all commercial power, and the renewable energy resources, primarily hydroelectric, is 6 percent. The United States is the world's largest oil importer, as we use foreign oil for 75 percent of our supply. Most of our imported oil comes from Canada followed by Saudi Arabia. The majority of the energy in the U.S. is used by industry.

Mining and smelting processes account for the majority of this energy use, followed by the chemical industry. Residential and commercial buildings use about 20 percent of the U.S. energy for heating, lighting, cooling, and water heating. Transport consumes 27 percent of the U.S. energy supplies in a year, with 98 percent of that energy supplied by petroleum.

Net energy production must be considered when evaluating energy resources. It takes tremendous amounts of energy to mine coal and transport the coal to a power plant. Seventy percent of the energy is lost in energy conversion at the power plant and 10 percent more during electrical transmission. Seventy-five percent of the energy lost as petroleum is converted into fuels, transported, and burned in vehicles. Natural gas has the greatest efficiency. It only loses about 10 percent of its original available energy and produces far less CO_2 per unit than coal or oil.

> **Take Note:** One of the 2005 essay questions on the AP exam addressed coal mine restoration and the problems associated with acid mine drainage. The question also expected the student to address other environmental costs associated with using coal for energy.

Coal

Coal is the fossilized remains of ancient plant material. Most coal was fossilized during the Carboniferous period, and therefore coal is considered a nonrenewable resource because it takes so long to form.

The first step in coal formation is peat. Peat bogs still exist throughout the world, and peat is burned as biomass fuel in many places. Peat produces less heat than any of the coals. As peat fossilizes, it forms the sedimentary rock coals. The softest coal is called lignite and it is found primarily in the western states. It has a lot of moisture in it and a woody texture. Bituminous coal is the most common coal and typically has a lot of sulfur. It has greater heat producing capacity than lignite. It is found in the Appalachians, Mississippi Valley, Central Texas, and the Great Lakes region in the United States. The coal with the greatest heat capacity is anthracite, a metamorphic rock. Anthracite is 95 percent carbon and has little sulfur, making it the cleanest burning of the coals. In the United States, the greatest reserves of anthracite are in Pennsylvania.

Coal Reserves

Coal is the most abundant of the fossil fuels. Proven reserves are stores of coal that can be removed with current technology at an economically feasible price. Known reserves are not fully examined, and the economic feasibility of its removal has not been established. The proven reserves of coal in the world should last approximately 200 years at the current rate of use. Coal reserves are not equally distributed around the world. The nations with the greatest coal reserves (in order) are the United States, Russia, China, India, and Australia. Coal is uncommon in Africa, the Middle East, and Central and South America.

Coal Mining and Environmental Damage

The two major types of coal mining in the United States are subsurface and surface mines. The subsurface, or underground mines, are found primarily in the eastern United States and are used for bituminous and anthracite coals. Subsurface mining is one of the most dangerous occupations in the world. In the past miners died from asphyxiation due to toxic gases, methane explosions, and cave-ins. Mines now employ exhaust fans to help prevent buildup of toxic gases. Miners also wear protective masks to prevent pneumoconiosis, also known as black lung disease. Black lung disease results from the breakdown of lung alveoli due to inhalation of coal particles. Gas exchange becomes inefficient and respiratory failure results. Some coal mines catch on fire and remain smoldering for years. One fire in China has been burning for 400 years, and one in Pennsylvania has been burning for over 40 years.

Subsurface mines may strike groundwater and contaminate it with heavy metals. Surface mines, often found in the western United States, are used primarily for lignite as it often lies fairly close to the earth's surface. Surface mines are also called strip mines, because the miners remove the topsoil in a strip of land and reserve it. They then remove the overburden with huge machines called draglines. The exposed coal seam is removed. A parallel strip is cut, and the overburden from the new strip is moved into the mined area and topsoil is replaced. The topsoil is replanted with native vegetation. The Surface Mining Control and Reclamation Act of 1977 (SMCRA) requires that the mined lands are reclaimed and restored by replacing the overburden and replanting the area. SMCRA also prevents mining in national parks, wild and scenic rivers, and national wildlife refuges.

A new type of mining, known as mountain top removal, has developed in the Appalachians. This mining involves removing the overburden from the top of a seam of coal using draglines and placing it in a nearby valley. The coal is then extracted from the mine. The entire topography of an area is altered when mountaintop removal is used. Miners apply for a variance to the Surface Mining Control and Reclamation Act to be able to carry out mountaintop removal.

Environmental Impact of Mining

There are numerous environmental impacts of mining. Mining destroys the natural vistas, creating unsightly scars on the earth's surface and destroying the aesthetic value of a region. Mining disturbs habitat for countless species and increases erosion. Erosion can be minimized by carrying out immediate reclamation. Mining may contaminate groundwater, particularly with acids or heavy metals. The iron pyrite frequently found in coal mines in the eastern United States dissolves in water and migrates into streams, acidifying those ecosystems. Underground mining generates tremendous amounts of solid wastes, known as tailings. These tailings are often contaminated with heavy metals.

Coal Combustion and Environmental Damage

Coal combustion emits more CO_2 per unit of heat than any other fuel. CO_2 is a greenhouse gas and contributes to global warming. Coal combustion also releases SO_2 and NO_x, which contribute to acid deposition. Coal combustion releases more nuclear radiation than any nuclear power plant. The particulates released form the fly ash that can contain heavy metals, such as arsenic, lead, cadmium, mercury, and zinc. Coal combustion is responsible for 25 percent of the mercury released in the United States. The bottom ash must be disposed of in a landfill. Coal burning power plants use the energy released from combustion to create steam to spin a turbine. To cool the water, a water source is required. Frequently lakes or the ocean are used, thus resulting in thermal pollution.

Methods to Remove SO_2 and Particulates From Coal Emissions

Sulfur dioxide may be reduced precombustion, during combustion, and postcombustion. Precombustion methods include using a higher grade of coal (anthracite) and washing the coal to remove excess sulfur. Coal may also be converted into a gas (coal gasification) or oil (coal liquefaction), which removes the sulfur. The process results in less net energy, because energy is lost as the solid is converted to a liquid or gas. A combustion method of removing sulfur from coal emissions is using fluidized bed combustion. Fluidized bed combustion is carried out by burning the crushed coal with crushed limestone. The sulfur in the coal combines with the calcium in the limestone to form calcium sulfate, or gypsum. This bottom ash can then be disposed of in a proper fashion. Postcombustion methods include using catalytic converters to oxidize the sulfur to yield sulfur compounds. A lime scrubber in a smokestack may also be used. In a wet scrubber, a slurry of lime mixed with water is sprayed across the exiting gases. The sulfur mixes with the calcium, forming the calcium sulfate, which falls to the bottom of the smokestack as bottom ash.

Particulates can be removed from coal combustion by burning coal with a low ash content. Most particulate removal is postcombustion. The resultant particle mixture must often be discarded in a hazardous waste landfill. Bag filters are a series of bags, somewhat like a bag in vacuum cleaner, which catch the particulates as they rise in the smoke. The bags are periodically emptied of their ash. Electrostatic precipitators remove 99 percent of the particulates in coal emissions. They function by passing the coal emissions past a series of charged plates, thus charging the particulates, which then bind to an oppositely charged plate. Cyclone collectors create a vortex in a smokestack, causing the particles to collide and fall to the bottom of the stack as bottom ash.

Oil

Oil is derived from ancient organisms buried in sediment and subjected to heat and pressure, which turned them into oil. Oil is usually found in conjunction with natural gas in porous rock, and they are held in place by impermeable sediments such as shale. In primary recovery, oil moves into wells, and it is pumped to the surface. By using secondary recovery, the yield is increased by injecting water into a second well to force oil toward the oil drilling well. Primary and secondary recovery allow 15-40 percent of a well's reserve to be removed. Tertiary recovery involves injecting steam or carbon dioxide into wells to stimulate the remaining heavy oil to flow to obtain another 5-15 percent of the oil remaining in the well.

Oil Reserves

Oil deposits have an uneven distribution throughout the world, but are present on every continent. Approximately two-thirds of proven oil reserves in the world are in the Middle East. The United States has about 3 percent of the world's proven reserves, primarily in the Gulf of Mexico, which provides 25 percent of U.S. oil and 20 percent of the natural gas. The oil in the Gulf is removed by deep wells situated below oil platforms, some stationary and some floating. The United States currently produces about 8 million barrels a day. As the price of oil continues to rise, current reserves considered too expensive to recover may become economically feasible.

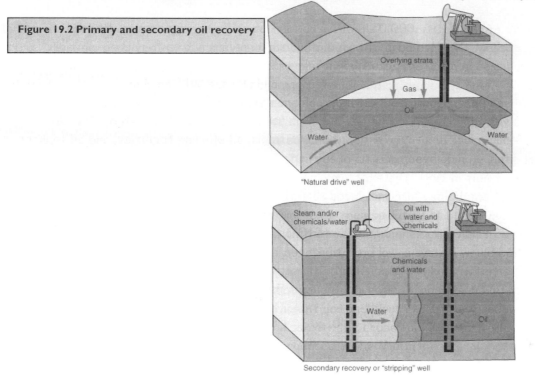

Figure 19.2 Primary and secondary oil recovery

Take Note: One of the 2005 essay questions on the AP exam addressed the controversial use of the oil lying beneath the Alaska Arctic National Wildlife Refuge (ANWR). Students had to be familiar with environmental damage associated with drilling for oil and how this environmental interference would cause damage to the tundra. Students were also expected to explain the uses of petroleum in the United States and propose mechanisms to decrease that usage.

ANWR Controversy

The coastal plain of the Alaska Arctic National Wildlife Refuge (ANWR) has a known reserve of oil that has not yet been tapped. The exact amount of oil in the plain is not known, but estimates run as high as 12 billion barrels. The plain is the calving ground of caribou and is the summering site for snow geese, swans, and numerous species of migratory waterfowl. In addition, polar bears, arctic foxes, and wolves are also found in this refuge. Preservationists worry that the traffic coming into the refuge may frighten the wildlife and cause them to leave. They say that improving the efficiency of internal combustion engines in cars will save for greater amounts of oil than ANWR will ever yield.   Proponents of the drilling plan claim there will be little damage done to the fragile ecosystem. Many people in Alaska favor drilling, because they work on rigs and receive oil royalties. Many in Florida consider ANWR a test case, to evaluate if Congress will place the country's need for oil above environmental preservation. Floridians have managed to hold at bay the oil companies desiring to drill off the coast of pristine Florida beaches. Many Floridians are concerned that opening up ANWR to oil exploration means that their highly desirable tourist beaches may soon be covered in tar and other debris from oil drilling.

Environmental Costs of Oil Recovery and Use

Oil tankers and pipelines leak, causing environmental damage. For example, in 1989 the EXXON Valdez ran aground releasing millions of gallons of crude oil in Prince William Sound, Alaska. To date it is the largest oil spill in the United States. Birds, sea otters, orcas, and salmon were killed. The salmon fishery was severely damaged that season, because few fish were harvested. The Oil Pollution Act of 1990 allows the EPA to better regulate spills, oil storage facilities, and oil tankers. One of the worst spills in history was deliberate, as the Iraqi army in Kuwait dumped a huge volume of oil into the Persian Gulf in 1991. The exact volume of oil will never be known, but is estimated to be many times that of the Valdez. Drilling leads to environmental damage including erosion and habitat disruption; however, little land space is used for wells and pipelines, compared to the vast amount of land disturbed by coal mining. The transport and building of the rigs and pipelines disrupts the habitat.

Petroleum is heated to remove the various components that will be used in a process called distillation or refining. Gasoline, kerosene, diesel fuel, lubricating oil, heavy gas oil, and asphalt are separated during the refining process. Products made from petrochemicals include plastics, paraffin wax, mineral oil, petroleum jelly, dyes, pesticides, and industrial solvents.

When oil or gasoline is burned, CO_2, a greenhouse gas, is released, which leads to global warming. When gasoline is burned, NO_x are released, which contributes to smog formation and acid deposition.

Tar Sand and Oil Shale

Oil can also be derived from tar sand or oil shale. Tar sands are sand or sandstone deposits infused with a thick oil known as bitumen. The oil is too thick to pump out in an oil well, so it must be heated to remove it from a well as a liquid or mined. The addition of steam to the tar sand deposits contaminates large amounts of water. The oil has to be refined to remove the sand. Several U.S. states have tar sand deposits, and Canada has a tremendous amount of this resource, primarily in Alberta.

Oil shale is a sedimentary rock composed of a variety of heavy oils called kerogen. The oil shale is surface mined, then crushed and heated to remove the oil. There are fairly large deposits of oil shale in the U.S., particularly in Colorado, Utah, and Wyoming. Its recovery is not yet economically feasible. The process also requires large amounts of water to refine the oil, but the resource in the United States is located in arid regions.

Natural Gas and Natural Gas Reserves

Ninety percent of the gas in natural gas is methane, followed by propane, ethane, and butane. Natural gas is the cleanest burning of the fossil fuels, and it has the highest net energy value. Natural gas pipelines run throughout the United States, and the United States has abundant reserves. To transport natural gas, it is usually cooled and compressed until it becomes a liquid. This liquid is known as liquefied natural gas, or LNG. In the United States, LNG is frequently used in rural areas. In urban areas, methane is piped to homes directly. Russia has slightly more than 30 percent of the world's known reserves of natural gas. Because the Middle East has so much oil, it is no surprise that they have 36 percent of the world's natural gas.

Unconventional Methane Stores

Methane can be found frozen in ice as methane hydrate. This methane can be removed by dissolving the ice in methanol. It is of great concern that if the ice caps melt due to global warming, the methane in the ice sheets will enter the atmosphere. Because methane is a greenhouse gas, global warming will be significantly exacerbated.

Methane has also been found associated with coal deposits in the western United States. This methane is relatively close to the surface of the earth, making its extraction economically and technically feasible. The methane is under groundwater supplies, which must be removed to extract the methane. This water tends to have large amounts of salt and other minerals and is thus too contaminated to release onto fields and pastures in the area. The water withdrawal is also drying up local wells that ranchers rely upon for their livestock. Livestock and wildlife are being killed by the traffic and waste left around the drill sites. The large numbers of wells are rapidly contributing to ecosystem disruption.

Take Note: There have been two essay questions on AP exams regarding nuclear energy. One was regarding Yucca Mountain and the disposal of radioactive isotopes. The other had students identify components of a nuclear power plant and explain their function. The question further asked students to identify problems associated with the use of nuclear power.

Nuclear Power

Despite all of the concerns, protests, and objections to U.S. nuclear power, the U.S. nuclear program has never had an accident at a plant where a significant amount of radiation was released into the environment. The Bush administration has renewed interest in the U.S. nuclear program, which had been waning in the last 25 years. Public concern, rising construction costs, and a declining demand for electricity have resulted in no new nuclear plants being built since 1975. Nuclear power is still nearly twice as expensive as coal energy. The 103 reactors in 31 U.S. states produce 20 percent of the country's electricity.

Nuclear Fuel Enrichment

Uranium exists in finite amounts and is therefore a nonrenewable resource. North America has roughly 22 percent of known uranium reserves in the world and Australia has 26 percent. The uranium must be mined, usually from sedimentary rock, which leads to all of the environmental impacts associated with mining a resource as discussed in chapter 14. The mine tailings are frequently radioactive and present additional dangers to the environment and surface water and groundwater supplies. Additionally, mine workers who work in uranium mines are susceptible to lung cancers induced by high radon levels in the mines. There are three isotopes of uranium. ^{238}U makes up 99.28 percent of all U in ore; ^{235}U makes up 0.71 percent, and ^{234}U makes up less than 0.01 percent. The desired isotope to use in a nuclear reactor is the fissionable ^{235}U, which must undergo a process known as enrichment to concentrate it to make up 3 percent or greater of the reactor fuel. The fuel is formed into pellets the width of a pencil and roughly one-inch long. The pellets are the equivalent of one ton of coal or four barrels of crude oil. There was an accident because of human error at an enrichment plant in 1999 in Tokaimura, Japan, which resulted in the deaths of two workers due to radiation exposure.

The pellets are placed in thin metal closed pipes approximately 12 to 16 feet long called fuel rods. These fuel rods are then bundled into clusters of 100-200 rods called fuel assemblies. A small reactor may have as few as 250 assemblies, but larger reactors can have up to 3,000 assemblies.

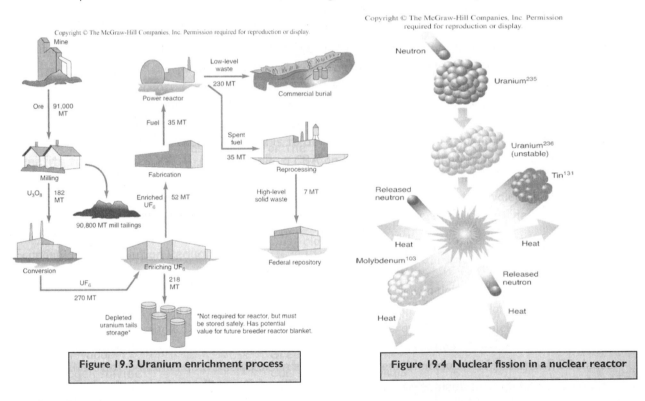

Figure 19.3 Uranium enrichment process

Figure 19.4 Nuclear fission in a nuclear reactor

Nuclear Fission

Nuclear fission is the splitting of an atom to release energy and particles. In nuclear reactors, the uranium is unstable, so when it is struck by a neutron, it splits into two smaller nonfissionable nuclei, releasing energy and more neutrons. These neutrons strike other nearby uranium atoms, initiating a chain reaction. To ensure that a chain reaction continues, the enrichment process increases the concentration of fissionable ^{235}U, and moderators are used. When the fuel rods are spent, they are replaced. Typically one-third of them are replaced at a time.

Moderators are needed because the rapidly moving neutrons created by the fission are readily caught by the ^{238}U still in the pellets. The fissionable ^{235}U is more likely to be struck by slower moving neutrons, therefore a moderator such as graphite, beryllium, H_2O (light water), or D_2O (heavy water containing deuterium [^2H] as the H, rather than hydrogen ^1H) is needed. Moderators slow rapidly moving electrons, which makes it more likely they will strike the ^{235}U.

Control rods are inserted in between the fuel rods to slow down a reaction or removed to increase the rate of fission. These control rods absorb neutrons, and therefore cadmium or boron is typically used.

Light Water Reactors

Seventy percent of nuclear reactors in the United States are pressurized-water reactors (PWR), with the rest boiling-water reactors (BWR). Both types are light water reactors, because they use light water (H_2O) as the moderator and coolant. In pressurized-water reactors, the energy from the nuclear reaction heats a primary water circuit that is under pressure and cannot boil. This water is piped to a secondary water circuit that does boil, creating steam to spin a turbine connected to an electrical generator. This secondary water circuit requires cooling to

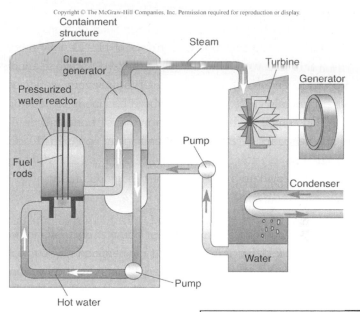

Figure 19.5 Light water reactor

be reused, and the tertiary water circuit is used for this purpose. The tertiary circuit uses surface water from a lake or the ocean or a cooling tower's water to pass around the secondary water circuit, removing the excess heat. The tertiary circuit then is cooled in the tower, or replaced in the surface water, resulting in thermal pollution of the surface water. In boiling-water reactors, the water in the primary circuit boils to spin the turbine to generate electricity. The steam is brought back to liquid state in a condenser cooled by a secondary water circuit.

The reactor vessel and the steam generator are contained within a stainless steel reactor vessel and a concrete containment building. The walls of the reactor vessel are up to 30 cm thick, and the reinforced concrete building is at least 1 m thick. The plants are constructed on geologically stable sites, and all plants in the United States have redundant safety measures and controls to prevent accidents. Many are concerned that terrorists will try to fly a plane into a nuclear plant. The air above a nuclear plant is considered a no-fly zone, with military planes at the ready should the airspace be breached. Additionally the plants are built to withstand damage, and the reinforced concrete and other safety features should withstand an attack.

It was discovered at a plant in Ohio that the boron in the control rods reacting with water in the reactor was forming boric acid and eating away at the surface of the stainless reactor vessel. Other plants have examined and rectified this problem as a result of the discovery

Other Reactor Designs

Canada employs heavy-water reactors, because they use D_2O as their coolant and moderator. These Canadian deuterium (CANDU) reactors do not require uranium enrichment to function. France, Great Britain, and the former USSR use graphite as the moderator and structural material for the core of the reactor. British reactors cool using CO_2 to cool the fuel assemblies and carry heat to generators. These graphite-based designs have always been considered safe, because graphite captures neutrons while dissipating heat. Unfortunately, as seen at Chernobyl, the graphite catches fire if the cooling system fails.

Breeder reactors create fuel rather than consume it. They create fissionable plutonium and thorium as the reactions within take place. Very few breeder reactors are currently in operation in the world due to safety concerns. The coolant is liquid sodium metal, which is highly unstable. The plutonium created could also be used in weapons, as it is removed from the reactor in a weapons grade state.

Numerous other plant designs are being tested throughout the world in an attempt to make nuclear power a safe, less air polluting alternative to fossil fuels.

Nuclear Accidents

Nuclear plants cannot explode as will a nuclear bomb because the uranium is not as highly enriched as in a bomb. The greatest danger in a nuclear plant is the loss of coolant. In this instance, the fuel will overheat, melting not only the fuel rods, but potentially the containment vessel. This is known as a "meltdown." The effect is the release of radioactive material. In 1979 in Three Mile Island, Pennsylvania, a partial meltdown occurred without gas release due to the secure containment building. The fuel rod meltdown was primarily due to worker error in conjunction with equipment problems. The coolant leaked out, resulting in uncontrolled fission. To date, no increased incidence of disease or death has been associated with this accident. The accident prompted stringent operator training, emergency response training, and radiation protection at U.S. nuclear plants. It also caused greater oversight by the Nuclear Regulatory Commission.

In April of 1986, reactor number 4 at the Chernobyl Nuclear Plant in the former USSR (now Ukraine) experienced a loss of coolant, resulting in the released steam blowing the roof off of the building and a severe graphite fire. Thirty people died immediately, and 135,000 had to be evacuated. Since, the thyroid cancer rate in children has increased dramatically. Americans assisted the Soviet government in encasing reactor 4 in concrete. This concrete has begun to fail, and private American companies are working with the Ukrainian government to decommission Chernobyl. Chernobyl lacked an appropriate containment building, which would have prevented the escape of the radioactive gases and fallout for the most part.

Radioactive Wastes

Until the 1970s, countries with nuclear power plants disposed of all wastes in the ocean. The Ukraine has dumped far more wastes than any other nation, primarily in the Arctic Ocean. They dumped nuclear reactors in the Kara Sea, many loaded with radioactive fuel.

Mining waste accounts for a large proportion of the waste generated by nuclear power. The tailings are left adjacent to mines, allowing water to transport the radiation into surface water and groundwater. Low-level radioactive waste, such as tools, building materials, contaminated clothing, cleaning materials, syringes, and other medical wastes, are stored in low-level radioactive sites. There are three such sites in the United States, located in Barnwell, South Carolina, Hanford, Washington, and Clive, Utah. The high-level wastes are highly radioactive. They are components of power plants including spent fuel rods. They are currently stored on the site of the power plants in deep pools of water. The ponds were intended to be temporary, but have now been used for over forty years due to the lack of federal facilities. Some facilities have begun to store wastes in above ground dry casks, but concern regarding the safety of this method of storage persists.

There are only three options when a nuclear plant is no longer useful. The plant can be encased in concrete and guarded for centuries. The plant may be placed under guard, and then slowly dismantled as the radiation threat decreases. The most dangerous option, decommissioning, involves tearing down a reactor while it is still highly radioactive. One great problem that exists is that the materials must be disposed of in a proper facility, which does not yet exist in the United States. One decommissioned plant in California still has no place to ship its waste, although Russia, notorious for dumping highly radioactive wastes into the Arctic Ocean, has offered to take radioactive wastes from other nations.

Highly contaminated materials must be stored in geologically stable areas not near water resources. Many countries use stable rock formations that occur naturally in the earth or in secure above-ground facilities. Several locations have been suggested to store these highly radioactive wastes, none of which are feasible. Outer space or shooting the wastes into the sun, are not practical because there is no way to ensure that the rocket carrying the waste will exit our atmosphere. A rocket loaded with waste that explodes in the atmosphere could contaminate the entire earth. Antarctica is not a viable option, because it is a pristine area and countries have agreed to protect it from human-induced damage. Many countries used to dump nuclear waste into the bottom of the ocean, but the possibility of corrosion of the vessels in the salt water and releasing highly contaminated waste is far too dangerous. It has also been suggested that the wastes be placed into subduction zones, but such wastes would be irretrievable and highly unstable due to the geologic instability.

Yucca Mountain

In 1982 Congress passed the Nuclear Waste Policy Act, which delineated that the federal government was responsible for developing a permanent site for highly contaminated radioactive wastes. In 1987 an amendment to the act identified Yucca Mountain, about 100 miles away from Las Vegas, Nevada, as the location. Yucca Mountain was supposed to be open years ago, but changing presidential administrations have resulted in fluctuating policy regarding opening the facility. It is predicted that the site will be ready to open by 2010. It is located on an active fault line and is near some ancient extinct volcanoes. The water table is far below Yucca Mountain and any accidental leaks are unlikely to result in groundwater contamination. That it is located in a desert eliminates precipitation carrying radioactivity to surface water. Although a large city is fairly near by, the state of Nevada has fairly low population density overall. Yucca Mountain was also used for early nuclear testing in the United States. The people in the state of Nevada are not overjoyed at having the radioactive waste from the entire country transported to their state. Grave concerns exist over the transport of these wastes from throughout the country to Yucca Mountain. An accident in the middle of a large city, for example, could be extraordinarily dangerous. Due to the spate of terrorist attacks, many feel that wastes are much safer in ponds next to nuclear plants than being transported via water, rail, or interstate. Others feel the process of transport will be safe, and that having the waste in a monitored, retrievable repository will be much safer than its current "temporary" storage.

Nuclear Fusion

Nuclear fusion could produce unlimited amounts of power; however, at this time, we do not know how to maintain a controlled nuclear fusion reaction. Humans have devised hydrogen bombs, uncontrolled fusion reactions. These reactions occur when nuclei of two atoms fuse releasing large amounts of energy. In the sun, atoms of deuterium and tritium or two atoms of deuterium fuse at temperatures over 100 million°C and pressures over several billion atmospheres.

Chapter 19 Questions

Use the following for questions 1-4.

    a. coal
    b. oil
    c. natural gas
    d. nuclear power
    e. methane hydrate

1. recovery of resource is not feasible with current technology and prices
2. combustion releases large amounts of mercury into the atmosphere
3. the most clean burning of the fossil fuels
4. must be distilled prior to use of the resource

5. Which of the following energy conversions is correct?
a. 1 BTU = energy required to heat 1 lb of water 1°F
b. 1 megawatt = 1,000 watts
c. 1 watt = 1 joule per second
d. 1 newton = force needed to accelerate a 1 lb mass 1 ft per second.
e. 1 kilowatt hour = number of megawatts used in an hour

6. The greatest proportion of U.S. electricity is provided by
a. coal power plants.   b. nuclear power plants.   c. oil burning power plants.
d. refuse derived power plants.   e. hydroelectric power.

7. Which of the following nations has the greatest supply of natural gas?
a. United States   b. Canada   c. Saudi Arabia   d. Russia   e. China

8. Which of the following is associated with mining uranium?
a. mesothelioma
b. pneumoconiosis
c. lung cancer
d. radiation sickness
e. gastrointestinal disease

9. Which of the following pollutants is not released by coal combustion?
a. sulfur dioxide
b. carbon dioxide
c. ozone
d. mercury
e. particulates

10. Coal mining has all of the following environmental effects except
a. increased erosion due to topsoil removal.
b. subsidence due to collapse of subsurface mines.
c. acid mine drainage in abandoned mines.
d. transport of coal results in damage to aquatic systems if tanker is damaged.
e. increased habitat disruption due to deforestation.

11. Which of the following structures in a pressurized water nuclear power plant is correctly paired with its function?
a. primary water circuit cools the fission reaction
b. secondary water circuit is heated and spins the turbine to generate electricity
c. water in the cooling tower is used to cool the primary water circuit
d. moderator absorbs neutrons to stop the fission reaction
e. containment vessel encompasses the electrical generator to prevent explosions

12. Which of the following is not an argument against drilling in the Arctic National Wildlife Refuge?
a. growing season is very short making it difficult for the ecosystem to recover
b. permafrost makes soil formation very difficult, resulting in a long succession time
c. area is a calving ground for caribou
d. high levels of biodiversity will allow the ecosystem to recover quickly
e. cold temperatures decrease the rate of nutrient cycling

13. A typical oil well can extract what percentage of the oil in the reserve by using primary recovery and secondary recovery?
a. 10
b. 30
c. 50
d. 80
e. 100

14. All of the following methods will reduce NO_x emissions from gasoline combustion except
a. catalytic converters in automobiles.
b. controlling the combustion temperatures.
c. switch fuel to ethanol or ethanol mixed fuel.
d. switch to a hybrid vehicle.
e. decreasing the release of ozone from auto emissions.

15. The fissionable component of nuclear fuel in a conventional nuclear power plant is
a. Uranium - 238.
b. Uranium – 235.
c. Uranium – 234.
d. Plutonium - 239.
e. Radon - 222.

Chapter 19 Answers

1. e. The recovery of methane hydrate is not feasible at current technology and prices.
2. a. The combustion of coal releases large amounts of mercury into the atmosphere.
3. c. Natural gas is the cleanest burning of the fossil fuels.
4. b. Oil must be distilled prior to use of the resource.
5. a. It takes 1 BTU to heat 1 lb of water 1°F.
6. a. The greatest proportion of U.S. electricity is provided by coal power plants.
7. d. Russia has the greatest supply of natural gas.
8. c. The disease associated with the mining of uranium is lung cancer due to radon exposure.
9. c. Ozone is a secondary air pollutant released when sunlight interacts with VOCs or NO_x.
10. d. Tankers leaking is a problem associated with oil transport.
11. b. The secondary water circuit is heated and spins the turbine to generate electricity.
12. d. The tundra has very little biodiversity, which will make it slow to recover.
13. b. A typical oil well can extract approximately 30 percent of the oil in the reserve by using primary recovery and secondary recovery.
14. e. Ozone is a secondary air pollutant resulting from NO_x emissions from gasoline combustion.
15. b. The fissionable component of nuclear fuel in a conventional nuclear power plant is Uranium – 235.

Chapter 20 - Sustainable Energy

Key Terms

active solar power   hydrogen fuel cell
biomass   methanol
energy efficiency   net energy yield
ethanol   ocean thermal electric conversion
geothermal energy   passive solar
green pricing   photovoltaic cells
hybrid vehicles   tidal station

Skills

1. Review methods of energy conservation.
2. Compare the different methods of transport that conserve energy.
3. Differentiate between types of alternative energy resources.
4. Contrast active and passive solar power.
5. Classify the different types of biomass fuels.
6. Identify methods to generate electricity from tides and ocean thermal gradients.
7. Examine the characteristics of geothermal power.

Take Note: Understanding energy conservation methods is extremely important for a student preparing for the AP Environmental Science exam. It is also of paramount importance that students are able to discuss the pros and cons of each type of alternative energy resource. Essay questions on past exams have addressed biomass fuel, water diversion projects, and wind energy. Questions regarding efficiency and alternative energy also occur on the multiple-choice portion of the exam.

Net Energy Yield

The net energy yield is the total useful energy derived from a resource minus the energy required to obtain the resource and make it available. Nuclear power has low net energy because it is expensive to mine and concentrate the uranium, build nuclear power plants, and dispose of the radioactive wastes. Coal has a high net energy yield because it is relatively inexpensive to mine coal and build coal burning power plants. The electricity generating renewable resource that has the highest net energy yield is hydroelectric power. Since passive solar power does not generate electricity, it has the highest net energy yield of all of alternative energy resources.

Electrical Grids

Most electrical plants are attached to a grid. Grids connect power plants to each other and to their customers. The peak demand is the amount of electricity needed during the times of day when electricity is most needed, such as late afternoon when people are arriving home and cooking dinner. A brownout occurs when the power is not sufficiently meeting the peak demand, but the grid does not completely fail. A blackout occurs when there is a malfunction in a power plant or in a grid. A rolling blackout occurs when areas lose power sequentially when demand is greatest. The power is then distributed over time if the power plants cannot meet the demand.

Cogeneration

Cogeneration is the simultaneous production of both electricity and steam in the same plant. Cogeneration is often used to be able to use waste heat from one process and use it in another process. For example, a steel plant using coal for heating metals could use the waste heat to heat buildings or even generate electricity. Many reverse osmosis water treatment facilities use cogeneration energy from a power plant to force the salty or brackish water through a semi-permeable membrane to produce fresh water.

Energy Conservation and Improved Efficiency

The easiest way to promote energy conservation is to promote energy efficiency. Energy efficiency is a measure of the energy produced compared to the energy consumed. The life cycle cost of an appliance or vehicle is the initial cost of the item plus its lifetime operating costs. Often energy efficient appliances cost more to purchase, but over the lifetime of the operation of the appliance will save money. For example, a new high-efficiency front loading washing machine uses far less water and electricity than a conventional top loading washing machine, saving the consumer those expenses. Additionally it spins clothing at such a rapid rate that drying time is decreased in the clothes dryer, further saving energy. The disadvantage is the initial cost, which may be significantly greater than a less efficient model. Therefore the life cycle cost of the front loader is far less than the life cycle cost of a conventional top loading machine. The biggest energy waster in most homes is the incandescent light bulb, only 5 percent efficient, thus losing 95 percent of its energy input. These bulbs can readily be replaced with a more expensive, compact fluorescent bulb that gives off four times as much light and actually lasts ten times longer than a traditional incandescent bulb.

Transportation Efficiency

Transportation can easily save energy by improving the fuel efficiency in vehicles. Automobile companies in the United States claim that it is too expensive to make automobiles more energy efficient, yet the energy efficiency standards mandated in the United States are far below those in other developed countries and even some developing countries. The use of hybrid vehicles, which have an internal combustion engine and a rechargeable battery to provide the extra energy required for speeding up and climbing hills, facilitates energy conservation. The battery in a hybrid vehicle may be charged by regenerative breaking or by plugging the battery into an electrical supply. Electric automobiles, promoted as energy efficient and nonpolluting, are only so if the source of the electricity used in the vehicles is renewable. Otherwise they simply shift a mobile point source of pollution to a stationary source, the power plant used to generate the electricity. Decreasing the number of automobiles on the roads by promoting carpooling, use of mass transit, walking, or bicycling also reduces energy expenditures and promotes conservation.

Home Conservation and Efficiency

Homes can easily be made more energy efficient. Improving building code standards in many areas would force homebuilders to improve the efficiency of homes. In addition to the change to energy-efficient appliances and lighting previously mentioned, home water heaters can be tankless, instant water heaters or changed to a more efficient natural gas water heater rather than an electrical water heater. Homes can be super-insulated, which means they have thicker insulation in their walls and attic. Homes so highly insulated do cost more to build, but the energy savings over time far outweigh the initial cost. Windows should be installed that have low emissivity (low E) to reduce energy requirements. These windows should face south to take advantage of passive solar heating. Straw bale houses, literally constructed from straw bales, are becoming more common. The houses must be covered with stucco to avoid rotting of the straw, but the houses are well insulated. Homes built into the sides of a hill, called earth-sheltered homes, use sod as their roof. The houses maintain a more constant temperature because the sod provides insulation, but require a well-reinforced roof due to the weight. Negawatt programs, where customers who increase their efficiencies receive rebates, are becoming more popular.

Sometimes simple measures can be taken to ensure that a home is more energy efficient. Homes should be kept airtight by plugging leaks around plumbing, doors, and windows to help prevent wasting energy. Keeping windows and doors shut when not in use saves energy because the outside air will not influence the inside temperature. Lights and appliances should be shut off when not in a room. Turning the thermostat up or down depending on the season will save energy. Attire can be modified to make the room temperature more tolerable. Increasing the efficiency of the air handling unit for the furnace/air conditioner will decrease energy waste, as does installing a programmable thermostat to ensure that temperatures are changed throughout the day if no one is at home. Darker roofs (lighter ones in hotter climates) will result in less energy cost as the heat from the sun is absorbed (reflected) to additionally warm (cool) a home. Operating ceiling fans in rooms (only when occupied) the rooms will feel cooler without having to increase the use of the air conditioner. Attic fans will reduce the amount of heat that builds up in the attic of a house, thus keeping the house much cooler.

When temperatures first begin to rise in the spring and wan again in the fall, opening windows on all sides of the house to allow airflow will avoid the need for central air conditioner use. Community education programs will assist the homeowner in making their home more energy efficient, as will rebates, tax write offs, or subsidies when purchasing energy-efficient appliances or increasing the amount of home insulation. The opposite could be true if individuals purchase more energy-inefficient models, by increasing the tax on these energy wasters.

> **Take Note:** It is important to separate energy efficiency from passive solar designs. A question on the 2006 AP exam involved explanation of passive solar designs, and many students focused on energy efficiency methods rather than correctly explaining passive solar energy.

Passive Solar Energy Designs

Passive solar heating designs capture the sun's energy in a building and then use the captured heat to warm the building as the heat is given off. Typically windows face south in passive solar designs to receive sunlight all day. The heat is absorbed by brick, stone, or adobe walls, which heat up and slowly release the heat during the night. An efficient passive solar system requires little to no supplementary space heating. Again, these homes cost a bit more to construct, but the life cycle cost is much less than a standard electrically heated home. These passive solar designs are most effective in sunny climates, but may be used in other places as long as another system is used to supplement the passive solar design on cloudy days. A relatively new passive solar design is to place a greenhouse on the south side of a building, which, with appropriate vents situated throughout the house, will heat the house the entire day. The flooring must be made of heat-absorbing material such as brick or stone. Trombe walls are walls made of heat-absorbing materials that have a layer of glass situated in front of them. The air within the space created by the wall and glass is heated and circulated by convection throughout the house. Trombe walls may also contain water, which circulates through the house to heat it.

Passive solar cookers, or solar box cookers, are small-scale ovens used to cook foods. They are a box lined with a reflective material that has a clear top. Although they can be used in developed countries, they are becoming more commonplace in developing countries. It may take more time to cook food, but women and children in developing countries do not have to scavenge all day for wood, nor do they have to spend hours inhaling particulates and other pollutants as they cook the food.

Cooling houses naturally can also be done easily, particularly in hot climates. By decreasing the amount of light entering the home, with shutters, curtains, shades, awnings, or even plants outside the windows, the house accumulates less heat and thus remains cooler. To further reduce heat buildup in the attic, installing lighter colored roofs will reflect more light, thus keeping the house cooler.

Active Solar Systems

In homes, active solar heating usually employs roof mounted solar collectors to heat water, air, or antifreeze. The collectors are usually black with layers of glass filled with the material to be heated and then piped directly throughout a home for heating. These systems may also be used to heat water for bathing and washing. This type of system is also common for heating swimming pools in colder months. Some of the solar collectors are passive in design as long as they do not employ a pump.

Photovoltaic (PV) cells are wafer thin sheets of silicon imbedded with boron impurities. When photons strike the glass plate, electrons are emitted from the wafer, creating a current. The cells were expensive in the past, but the cost of the cells continues to drop. Many cells joined together are useful to generate electricity. The cells can be joined in any number and any array to generate the amount of electricity desired. Although making the PV cells pollutes water, no emissions result from the function of the PV cells. Some consider PVs aesthetic or visual pollution because they are visually unappealing, thus affecting the aesthetic value of a house or business. A new type of technology, called an amorphous silicon collector, is currently being used in calculators, watches, and toys and is being developed as a new type of PV cell.

Commercial Solar Designs

Solar thermal electricity is derived when the sun's rays are focused on a system filled with a heat absorbing liquid. The liquid then heats water, creating steam to spin a turbine. Solar power towers are systems of mirrors built around a tower. Remote-controlled sensors shift the mirrors to ensure maximum sun exposure and thus reflection onto the tower. The liquid in the tower is usually a molten salt, which then heats water to create steam to spin the turbine to create the electricity. The salt holds heat in for quite some time, and therefore the energy can even be used at night.

Pros and Cons of Solar Power

The environmental benefits to using solar energy are that the fuel is free and no pollutants are emitted as the energy is used. The active solar systems are flexible, in that a small system may be developed or a large system may be developed. Drawbacks to using solar power include that the source (the sun) is intermittent in many locations and thus this type of power is best suited to relatively sunny climates. The ability to store solar heat is limited, so using solar power to generate electricity will require more advanced technology in the future. Lead acid batteries cannot store large amounts of energy per unit mass. Acid is corrosive and the lead is a heavy metal, which must be mined and smelted. Other battery types are equally problematic. Active solar designs can be expensive at first, and commercial operations require large amounts of space. Large systems may require vast amounts of land, which may result in deforestation, erosion, and habitat disruption depending upon the location of the system. However, the sites may also be used as rangeland, so it does not have to be fully devoted to electrical generation. The energy created by solar power is direct current (DC), so unless direct current appliances are present, alternating current (AC) converters must be used.

Wind Energy

For centuries wind has been used in windmills to pump water and perform mechanical processes such as grinding grain. Present-day mechanical turbines use long propeller blades that generate large amounts of torque, which generates more electricity than a faster spinning windmill with shorter blades. Groups of wind turbines are known as wind farms. Offshore windmills are becoming a new resource of reliable energy. Frequently these structures can be placed far enough offshore that it does not disturb the aesthetics of a shoreline. A large scale offshore wind farm has been planned off the coast of Martha's Vineyard and Cape Cod, but locals are concerned about impaired boat traffic and the natural landscape.

There are numerous benefits to using wind energy. The source of the energy is free, and the windmills emit no pollutants as they function. Wind is unlimited at favorable sites. The disadvantages of using windmills include that they are unsightly and impair the aesthetic value of an area. The wind is also an intermittent source, and thus wind farms are usually strategically placed in areas where wind is relatively constant. The wind turbines are often noisy and may affect animal behavior in an area. Birds, particularly migratory birds and raptors, have difficulty seeing the propellers and are killed. These bird deaths can be avoided by adding antiperching devices, painting the blades a color the birds can see, and having the blades make noises that irritate the birds. Population density is somewhat low in areas with high wind, so power lines may need to be put into place to convey the power to an area that needs it. Also, wind power is difficult to store and would require battery storage.

Hydroelectric Power

Hydropower functions by building dams across rivers and streams. 20 percent of the commercial energy in the world is derived from falling water. The largest hydroelectric dam in the United States is the Hoover Dam, and the largest in the world will be the Three Gorges Dam on the Yangtze River. Water is released through pipes in the dam that have turbines in them to spin a generator to make electricity. Hydroelectric power does not generate pollution as the dam functions. The net energy yield is also the highest of all of the alternative energy resources that generate electricity. The review of chapter 17 discusses the pros and cons of water diversions.

Smaller dams may also be built to generate electricity and decrease the environmental impact of the dams. These dams include low-head hydropower dams built on headwater streams and micro-hydro generators that power individual homes.

Geothermal Power

Geothermal power is derived from the energy produced as radioisotopes decay in the earth's surface. This decay heats rock or groundwater (geysers or hot springs) that can be used to generate electricity. The steam generated by the heated water then spins a turbine connected to an electrical generator to create electricity. The geothermal plants have a long lifespan and do not require land degradation for mining of materials. There are few wastes that need disposal. The water tends to have a lot of salts associated with them, so corrosion of power plant equipment may occur. When too much groundwater is used, the surrounding land may subside. Examples of areas that use geothermal power include Iceland; Boise, Idaho; and The Geysers, California.

Wave and Tide Power

Waves are used to generate electricity when they pass through a turbine causing it to spin. There are only a few commercial wave generating facilities throughout the world. The first was built in Scotland in 2001. The technology to harness wave energy is still being researched, as are the system designs that will withstand adverse weather conditions.

Tidal power is usually most efficiently generated when there is a large difference between high and low tide. An example of such a locale is Annapolis Royal in Nova Scotia. When the tide comes in, the water spins the turbine in a tidal station creating electricity. When the tide goes out, the turbines also spin, which allows continued generation of electricity. One problem with tidal power is the damage that occurs from building dams across inlets and bays. The greatest tidal energy is usually found in estuaries, and damming these areas can result in ecosystem damage including siltation and damage to breeding areas.

Ocean thermal energy conversion (OTEC) systems are designed to use the thermal gradients that are present in water columns to generate electricity. These systems run fluid through the layers of ocean water to heat and then cool it. When the fluid is heated, it spins a turbine generating electricity. The best sites for this type of energy are along shorelines of volcanic islands, because their temperature gradients are more distinct than in other parts of the ocean.

Biomass Fuels

Biomass fuels are any organic matter that can be burned. Most biomass fuels are created via photosynthesis, including wood, charcoal, peat, and crop residues such as cornstalks, corncobs, or wheat straw. There are numerous problems associated with using wood and charcoal for energy. Wood smoke has numerous pollutants, especially particulates and carbon monoxide. The temperature of combustion may be controlled and emission control devices placed on chimneys to decrease air pollution from burning wood. Many developing countries are experiencing a wood shortage. It is estimated that 1.5 billion people lack enough wood or charcoal to heat their homes and cook their food.   Women and children spend hours searching for wood. Previously densely wooded areas are now barren due to the deforestation resulting from forests being cut down to acquire the wood or charcoal necessary for life.

Methanol and ethanol can be used in internal combustion engines. Grains, sugar cane residue, or sugar beets may be used to generate ethanol. Methanol may be derived from wood, wood waste, crop residues, and sewage sludge. These biofuels typically have a high net energy yield and reduce agricultural wastes. They do require large amounts of land, and fertilizers, pesticides, and fossil fuels may be used in large amounts to generate the biomass crops.

Biogas digesters, which employ anaerobic bacteria to digest wastes, can decompose household or animal feces to create methane. In developing countries, these feces are also in demand as fertilizers, and thus may limit agricultural capability resulting in a lower food supply. The methane combustion creates far more heat than directly burning feces. Dung can also be burned directly once it is thoroughly dried. Many municipal sewage treatment plants use anaerobic digestion during their sewage treatment and use the methane produced to run their operations.

Hydrogen Power

Hydrogen burns readily and when combined with atmospheric oxygen, creates water vapor. Some NO_x and CO_x are also formed as a byproduct. Hydrogen is removed from a fossil or biomass fuel using a reformer in conjunction with a fuel cell, but researchers seek to find a renewable source of hydrogen. There are numerous types of fuel cells, but all contain a catalyst and an electrolyte solution to generate an electron stream for generation of electricity. For example, plants use sunlight to split water to form hydrogen and oxygen, and researchers would like to be able to use water in a similar fashion.

Chapter 20 Questions

Use the following for questions 1-4.

    a. geothermal energy
    b. passive solar power
    c. active solar power
    d. hydroelectric power
    e. biomass power

1. uses the natural decay of radioactive isotopes in the earth's crust to heat water to spin turbines
2. exemplified by photovoltaic cells
3. examples include dung, peat, and ethanol
4. has the greatest net energy of all of the alternative energy resources

5. If your laptop computer uses 50 watts per hour and you use it for three hours per day, how much will the electricity cost to run the computer for one year if your utility charges $ 0.08 per kilowatt (kWh) hour?
a. $10.46   b. $8.52   c. $5.00   d. $4.38   e. $2.98

6. Which of the following are environmental costs associated with the generation of geothermal power?
            I. land subsidence
            II. groundwater depletion
            III. carbon dioxide emissions
a. I only   b. II only   c. I and II   d. I and III   e. I, II and III

7. Which of the following is a passive solar design?
a. increased insulation to keep warm air in a home in the winter
b. using reflective roofing to decrease cooling costs in the summer
c. using photovoltaic cells to generate electricity
d. installing energy-efficient windows to keep hot air out in the summer to decrease air conditioning costs
e. planting a tree line of conifers to block wind from reaching a house in northern climes, which decreases winter heating costs

8. Which of the following are associated with the use of hydroelectric dams?
            I. increased temperature downstream
            II. decreased dissolved oxygen downstream
            III. increased sediment downstream
a. I only   b. II only   c. III only   d. I and II   e. I, II, and III

9. All of the following are biomass sources of energy except
a. methanol.   b. wood.   c. crop residues.   d. lignite.   e. charcoal.

10. Which of the following are associated with commercial active solar power generation?
    I. little land use and habitat disruption
    II. mining silicates for PV cells results in erosion
    III. thermal pollution from cooling power plants

a. I only   b. II only   c. III only   d. I and II   e. I, II, and III

11. Which of the following has the greatest net energy for heating homes?
a. coal   b. nuclear   c. hydroelectric   d. biomass   e. natural gas

12. Which of the following are problems associated with wind power?
    I. bird deaths
    II. aesthetic pollution
    III. noise pollution

a. I only   b. II only   c. III only   d. I and II   e. I, II, and III

Answers Chapter 20

1. a. Geothermal energy uses the natural decay of radioactive isotopes in the earth's crust to heat water to spin turbines.
2. c. Active solar power is exemplified by photovoltaic cells.
3. e. Examples of biomass power include dung, peat, and ethanol.
4. b. Passive solar power has the greatest net energy of all of the alternative energy resources.
5. d. It would cost $4.38 for one year.
   50 w x 3 hr/day x 365 days/year x kw/1,000 w x $.08/kWh = $4.38/year
6. c. Land subsidence and groundwater depletion are some environmental costs associated with the generation of geothermal power.
7. b. A passive solar design is using reflective roofing to decrease cooling costs in the summer. Increased insulation to keep warm air in a home in the winter, installing energy-efficient windows to keep hot air out in the summer to decrease air conditioning costs, and planting a tree line of conifers to block wind from reaching a house in northern climes decreasing winter heating costs are energy-efficiency methods. Using photovoltaic cells to generate electricity is an active solar design.
8. e. All are associated with the use of hydroelectric dams.
9. d. Lignite is a type of coal.
10. b. Mining silicates does result in erosion. A lot of land is used in commercial operations but there is no thermal pollution.
11. a. Coal has the greatest net energy for heating homes.
12. e. All are problems associated with wind power.

Chapter 21 – Solid, Toxic, and Hazardous Waste

Key Terms

biodegradable plastic	hazardous waste	secure landfill
bioremediation	incineration	Superfund
brownfields	photodegradable plastic	Toxic Release Inventory
composting	phytoremediation	waste to energy
demanufacture	recycle	
e-waste	sanitary landfill	

Skills

1. Compare a sanitary landfill with an open dump.
2. Examine the costs and benefits of recycling wastes.
3. Differentiate between the terms reduce, reuse, and recycle.
4. Analyze the costs and benefits associated with incineration.
5. Define hazardous wastes. Explain how such wastes are disposed of in the United States.

> **Take Note:** Essays on past AP exams have asked students to address pros and cons of recycling. The questions have addressed paper and aluminum recycling, in particular, including recommending specific policies to encourage recycling. On the 2006 exam, one question addressed brownfields and the issues associated with reducing and disposal of hazardous wastes. You must be familiar with the methods used to dispose of wastes and ways to reduce waste. In addition, you must be familiar with what wastes are deemed hazardous and how those wastes must be handled differently from municipal solid wastes.

Solid Waste

Any waste that is not a liquid or gas is considered to be solid waste. The greatest sources of solid waste in the United States are mining and agriculture. Mining waste includes tailings, overburden, and smelter waste. Agricultural waste includes crop residues and animal waste. Other industries contribute to solid waste as well. Municipal solid waste (MSW) makes up a relatively small proportion of the solid waste produced in a country. The components most commonly found in municipal solid wastes are paper and paper products, yard trimmings, food scraps, plastics, metals, textiles, wood, and glass. Frequently household waste contains hazardous materials including paint, cleaners, oils, batteries, or pesticides.

Waste Disposal Methods

In many countries, an open dump is used for waste disposal. They are unsanitary, dangerous sites in which municipal waste is often not separated from hazardous wastes. The dumps are malodorous, vermin-infested sites that tend to catch fire, explode, and collapse under unsuspecting trash scavengers. The fires and explosions are caused by decomposition, which produces the flammable gas methane. Many disenfranchised poor live on or near these dumps, because they can find food, clothing, and shelter components in the trash. Although developed countries have modern sanitary landfills, illegal dumping on roadsides and vacant lots continues to be a problem. A sanitary landfill is designed to prevent vermin and spread of disease while providing trash disposal. The landfill is lined with layers of clay and plastic to decrease the amount of fluid, called leachate, which would flow out of the landfill. The leachate is collected in a series of pipes and treated as wastewater prior to being released. The trash in the landfill is compacted and covered with a layer of soil before more trash is placed on top of it. There are also pipes that traverse the landfill that collect methane. This methane can be used as a source of energy or burned off. Landfill sites must be geologically stable areas, and should be relatively close to the municipality it serves to save on transportation costs. Most people do not want a landfill in their vicinity, so (Not In My Backyard) is a very prominent factor when building a new facility.

Toxic colonialism is sending solid and hazardous wastes to Third World countries to dispose of it, which exposes the citizens of that country to hazards they would not expect. Industrialized countries have agreed to ban this practice, but illegal dumping continues. The Basel Convention of 1989 banned trade in hazardous waste from developed to developing countries. The United States failed to ratify the Basel Convention and exports much of our e-waste to Asia. E-waste is comprised of discarded computers, printers, cell phones and other technology waste. Additionally, poor neighborhoods and Native American reservations are subject to toxic colonialism because they do not have the wealth or the support to fight waste disposal site development on their land.

Ocean dumping is illegal in the United States due to the Ocean Dumping Ban Act of 1988. The act specifies that sewage sludge, industrial waste, medical wastes, or municipal solid wastes may not be dumped into the ocean. Dredge spoil, produced when channels are dredged to deepen them, are still dumped at sea. These wastes may contain heavy metals, pesticide residue, and other chlorinated hydrocarbons such as PCBs.

Incineration

Incineration is another method to dispose of solid waste. The process of burning the waste reduces the amount of landfill volume used because only the ash remains to discard. The landfill volume is reduced to 10-20 percent of the original landfill space. Many incinerators remove waste that will not burn, or that may be recycled, prior to combustion. These items include metals, glass, and other noncombustible materials. The trash has been enriched to readily burn and is known as refuse derived fuel. Other incinerators can burn any trash smaller than a refrigerator and are known as mass burn incinerators. These incinerators save money because they do not sort the different noncombustible components. The cost of an incinerator is high, and frequently garbage disposal in areas using incinerators is more expensive.

When waste is burned, it releases many air pollutants, including dioxins, furans, lead, mercury, and cadmium. These toxins are more prevalent in the fly ash than in the bottom ash, thus requiring postcombustion control mechanisms to prevent their release. Both electrostatic precipitators and bag house filters used to control emissions from fossil fuel combustion will function to reduce release of these toxins in incinerator emissions. Precombustion methods may also be used, including removal of batteries and chlorine-containing plastics prior to burning the refuse.

Some incinerators are waste-to-energy or energy recovery units. These units generate electricity by burning the trash to heat water, which becomes steam. The steam spins a turbine connected to an electrical generator. The steam may also be used to directly heat a building.

Some incinerator ash is combined with concrete or asphalt to create road fill. Many people are concerned that hazardous materials in the ash may be released as the road wears away. Incinerator ash that contains large amounts of hazardous materials must be disposed of in a hazardous waste landfill.

Dioxins

Dioxins are a group of chlorinated hydrocarbon chemicals. The most toxic is 2, 3, 7, 8-tetrachlorodibenzo-p-dioxin or TCDD. They are naturally produced during forest fires but the anthropogenic sources include incineration, smelters, chlorine bleaching at paper mills, and environmental tobacco smoke. Humans are exposed to dioxins when they ingest contaminated animal fat, because the chemicals bioaccumulate in fat and biomagnify. They are highly persistent. Acute exposure in humans results in chloracne, skin rashes, and skin discoloration. Chronic exposure results in liver damage. Dioxins are known to be teratogenic and immunotoxic and are thought to be a human carcinogen.

Reducing Waste

The three Rs apply when trying to reduce the volume of waste. Reduce, reuse, and recycle are the primary ways to reduce waste, and they are preferred in that order. Reduction is to decrease the amount of wastes by decreasing packaging, or redesigning a product to use less material, which is called source reduction. Source reduction is lowering the amount of material in a product. For example, aluminum cans use much less aluminum now than in the past. Reusing is to use the waste for another or the same purpose. For example, reusing a plastic grocery sack as a garbage bag decreases total waste volume. Reusing glass beverage containers, such as a glass soda bottle, reduces the amount of glass thrown away. A glass bottle can be reused an average of 15 times before it is too damaged to be used any longer.

Recycling also reduces waste by converting the waste into another product or the same product again after processing. Open-loop recycling is converting a product into a different product. For example, newspaper may be remade into another paper product such as notebook paper. Closed-loop recycling is employed when a product may be recycled into the same product. For example, it is estimated that a soda can may be recycled into a new can in less than three months time.

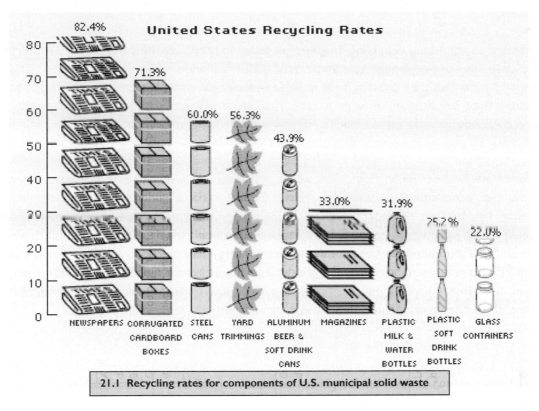

21.1 Recycling rates for components of U.S. municipal solid waste

To create a market for recycled material, it must be economically profitable to recycle. Recycling should also save raw materials and reduce energy use, and landfill space—in addition to being more economical than using the virgin resource. Recycling should also reduce air pollution. One of the most successful recycling programs is the recycling of steel. Nearly all steel in the United States contains at least 25 percent recycled steel and some contain 100 percent recycled steel. Much of this steel comes from steel cans, automobiles, buildings and bridges, and appliances. Mining and smelting iron to obtain steel consumes far more energy than melting old steel and forming new steel cans or girders. Aluminum, derived from an ore called bauxite, must be mined and extracted from the ore in a complicated process. By recycling aluminum in cans, the costs associated with the extraction of virgin ore are dramatically lowered. Aluminum is also easy to recycle and can easily be melted and remade into a can almost immediately. Approximately two-thirds of all aluminum cans are now recycled. Other metals that may be recycled include lead (auto batteries), platinum (catalytic converters), gold, silver, copper (pipes and electrical wiring), and iron.

Lead and Lead Toxicity

Lead is easily recycled from used automobile batteries. Lead is of great concern because it bioaccumulates in bone, unlike most toxins, which bioaccumulate in fat. Lead toxicity, particularly in the young, leads to mental retardation, lowered IQs, hyperactivity, and attention deficit and learning disorders. In the United States the maximum level of lead has been established at 10μg/dL blood. Atmospheric sources of lead include leaded gasoline (banned in the United States in 1976), smelters, and incinerators. Lead also contaminates soil and water from improperly treated leachate, lead pipes, lead solder, and leaded paint.

Paper Recycling

Paper is exposed to open-loop recycling, because as paper is recycled the fibers become shorter and shorter until they are no longer able to be used. Paper is made primarily from trees. The tree bark is stripped from the tree and the tree is pulped into small chips. Chemicals are added to the chips to ensure their breakdown. A large volume of water is required to repeatedly rinse the pulp. The paper is bleached, usually using chlorine. Other bleaching agents include hydrogen peroxide and oxygen gas.

Recycling paper involves de-inking first. The waste paper is usually mixed with fresh pulp, because the fibers shorten each time the paper is recycled. For example, old newspaper is usually recycled into newspaper, egg cartons, or paperboard. Recycled paper requires the use of less chlorine and water to create than does virgin paper. Virgin paper production usually results in air contamination, particularly sulfur dioxide and particulates from the burning of fossil fuels to operate the plant. These air pollutants are reduced in recycling facilities, because it takes 20 percent less energy to recycle paper than to make virgin paper. Recycled paper requires that fewer trees be harvested, thus reducing all of the environmental impacts associated with deforestation.

Composting

Yard waste is an unusual problem. It makes up such a large proportion of U.S. municipal solid waste, so many communities have made it illegal to dispose of yard wastes in their trash. Instead, these communities have established area composting facilities, in which the yard waste and tree trimmings are dropped at a central location to be pulverized into mulch. This mulch can then be used as a green fertilizer. Mulch not only increases the nutrients in soil, it improves the water holding capacity and oxygenation of the soil, in addition to preventing erosion.

Recycling Difficulties

Plastics can be difficult to recycle. The plastics in a soda bottle, egg carton, garbage bag, and a shampoo bottle are all different compositions. Soda bottles are made of a plastic called PET, which can be recycled into carpets, clothing, bottles, and packaging. Contamination with a minuscule amount of PVC plastic can render PET unusable for recycling. Recycled plastic is also more expensive than virgin plastic, due to the low price of petroleum. Mixed plastics can be made into toys and plastic lumber.

Some plastics are marketed as biodegradable, as they contain some materials that can be broken down by bacteria. These products are only partially degradable, because the plastic component still cannot be degraded. Other plastics are photodegradable, which means they are broken down by exposure to light. The obvious problem in this case is that in a sanitary landfill, light penetration is highly unlikely, and these plastics never degrade.

Tires cannot be easily recycled, although they can be reused. Tires are made of vulcanized rubber, which cannot be remelted and used as rubber again.   Tires tend to rise in landfills, which make them difficult to bury. Shredded tires fit more easily into landfills. These tire shreds can be used on playgrounds and as artificial mulch. Tires may also be burned in an incinerator to produce energy.

Demanufacturing

Demanufacturing is taking apart complex household items and retrieving the recyclable components. Items subject to demanufacturing include refrigerators, stoves, televisions, and air conditioners. Personal computers contain heavy metals, plastics, and valuable metals that can be removed and recycled as previously noted. Eighty percent of the e-waste produced in the United States is sent to Asia for disposal, which is legal because the United States never ratified the Basel Convention.

Hazardous Wastes

Hazardous wastes include any wastes that are flammable, explosive, corrosive, or highly reactive. In addition, chemicals that are toxic, or that induce cancer, mutation, or birth defects in living organisms are considered hazardous. Hazardous wastes do not include radioactive wastes. Much of the hazardous waste produced in the United States is recycled, stored, or converted into a less hazardous material.

The easiest way to control hazardous wastes is to avoid using hazardous materials in manufacturing processes. The wastes can also be reused and recycled during industrial processes, which will decrease the amount of wastes. Green, or environmental chemistry, is the study of chemistry to redesign chemical processes to be less hazardous.   Some physical processes to remove hazardous material include using charcoal to absorb toxins and distilling hazardous chemicals out of aqueous solutions. Some chemicals, such as PCBs, may be destroyed by high-temperature incineration. However, this procedure is difficult when soil or water is involved. Incineration may release dioxins and create an ash that must be permanently stored. The ash is usually much less toxic than the original waste. Long-term storage is also an option. It involves encasing the chemicals in noncorroding containers to be stored for long periods. The storage is in specially modified landfills called secure landfills. The leachate from these sites is carefully monitored. Permanent retrievable storage is keeping the wastes in a secure location where they can be inspected periodically and retrieved as needed.

Bioremediation is using bacteria and other microbes to break down hazardous wastes. Bacteria have been found, or genetically engineered, that can degrade chemicals such as PCBs, organic solvents, and pesticides. Phytoremediation is using plants, which may be genetically engineered, to absorb and accumulate toxic material from the soil. For example, sunflowers remove lead, poplar trees remove many contaminants, and canola removes selenium. These plants become contaminated with the material and must be disposed of properly. Phytoremediation is a slow process and works only as deeply as the roots of the plant can penetrate the soil.

The Problem of PCBs

Polychlorinated biphenyls, or PCBs, are a group of chlorinated hydrocarbons that are unusually persistent and bioaccumulate in fat. They also biomagnify in the food chain. The chemicals were made in the United States until 1976 and were used in electrical transformers, capacitors, vacuum pumps, and turbines. The liquid PCBs were used as adhesives, lubricants, fire retardants, and hydraulic fluids. Acute exposure to PCBs causes nausea, diarrhea, and vomiting. Chronic exposure interferes with the endocrine system and may cause cancers. PCBs can be removed by bioremediation and destroyed by high-temperature incineration.

RCRA

The main legislation that controls hazardous waste disposal is the Resource Conservation and Recovery Act (RCRA). This act was passed in 1976 to ensure testing and correct disposal methods for all hazardous materials. The legislation requires cradle-to-grave tracking of all hazardous material used in the United States.

Love Canal

Love Canal, New York, has the dubious honor of being the first location in the United States to be declared a national emergency disaster. The entire community was evacuated once it was realized the inhabitants were living on a site filled with toxins and carcinogens. A chemical company had been storing hazardous wastes in a canal for decades. The dump site was closed and a neighborhood built upon it. After a particularly rainy year, the chemical storage tanks began leaking their toxic materials. Children began getting chemical burns and the area smelled like chemicals. The families suffered numerous miscarriages and birth defects. After years of litigation, the chemical company was required to clean up the site. The situation at Love Canal was the impetus for passing the Superfund Act.

CERCLA and SARA

The Comprehensive Emergency Response, Compensation, and Liability Act (CERCLA), also known as the Superfund Act was ratified by congress in 1980. In 1984 the act was amended by the Superfund Amendments and Reauthorization Act (SARA). The acts are intended to allow the federal government to respond quickly to hazardous waste site contamination. The EPA administers the act and determines which sites require immediate attention due to supertoxic, carcinogenic, mutagenic, or teratogenic chemicals. The chemicals of greatest health concern on Superfund sites are lead, trichloroethylene, toluene, benzene, PCBs, chloroform, phenol, arsenic, cadmium, and chromium. These sites are placed upon the National Priorities List. There are currently 1,244 sites on the National Priorities List. The sites include abandoned factories, smelters, mills, refineries, and chemical plants. If the government cannot establish who is responsible for the damage, the cleanup is paid for from the Superfund. The Superfund money has been generated by taxing chemical and hazardous materials. The tax expired in 1995 and has not been reauthorized. Current funding comes from public funds. Companies are required to report their Toxic Release Inventory, a report on the hazardous materials they release in a year.

Brownfields

Brownfields are sites contaminated with toxic or hazardous materials. These properties have been abandoned and are not contaminated enough to be placed on the National Priorities List. Many industrial areas in the interior of urban areas are brownfields. These areas are subject to remediation to allow the area to be reclaimed and repopulated by humans. These reclaimed areas can provide jobs, a tax base, and housing in urban areas previously unavailable. Urban sprawl is curtailed when brownfields are used.

Chapter 21 Questions

Use the following for questions 1-4.

a. phytoremediation
b. photodegradable
c. composting
d. incineration
e. sanitary landfill

1. exemplified by sunflowers removing lead from contaminated soil
2. exemplified by household food waste decomposing into a soil conditioner
3. used to dispose of more than 50 percent of the U.S. solid waste
4. exemplified by plastics that break down upon exposure to light

5. Refuse derived fuel
a. is trash that has not been sorted.
b. consists primarily of agricultural wastes.
c. is used in mass burn incinerators.
d. is garbage sorted to remove noncombustible or recyclable materials.
e. is used to reduce the toxic materials in hazardous wastes.

6. All of the following are metals released from incinerators except
a. lead.   b. mercury.   c. dioxins.   d. cadmium.   e. copper.

7. Place the following terms in order with regard to saving energy (lowest energy first).
a. reuse, reduce, recycle   b. reduce, reuse, recycle
c. reduce, recyle, reuse   d. reuse, recyle, reduce
e. recyle, reuse, reduce

8. Which of the following is not an environmental consideration when recycling aluminum?
a. erosion caused by mining bauxite
b. cost of recycled aluminum compared to virgin aluminum
c. landfill space saved by recycling aluminum
d. less energy required to recycle aluminum compared to energy required to smelt ore
e. damage to habitats caused by mining of aluminum ore

9. The Superfund
a. is used only to clean up hazardous waste sites owned by the Department of Defense (DOD).
b. pays companies to clean up the land that the company accidentally contaminated.
c. identifies the worst hazardous waste sites and places them on the National Priorities List.
d. regulates the closure of sanitary landfills.
e. promotes conversion of open dumps to sanitary landfills in developing countries.

10. All of the following would be considered brownfields except
a. an abandoned fertilizer plant.
b. an obsolete closed chemical plant.
c. a closed smelter.
d. an abandoned dry cleaning facility.
e. a closed sanitary landfill.

Chapter 21 Answers
1. a. Phytoremediation is exemplified by sunflowers removing lead from contaminated soil.
2. c. Composting is exemplified by household food waste decomposing into a soil conditioner.
3. e. Sanitary landfills are used to dispose of more than 50 percent of the U.S. solid waste.
4. b. Photodegradable plastics break down when exposed to light.
5. d. Refuse derived fuel is garbage sorted to remove noncombustible or recyclable materials.
6. c. Dioxins are not metals.
7. b. reduce, reuse, recycle
8. b. The cost of recycled aluminum compared to virgin aluminum is an economic consideration when recycling aluminum.
9. c. The Superfund identifies the worst hazardous waste sites and places them on the National Priorities List. It is not only used to clean up DOD sites. One goal of the EPA is to ensure that companies that contaminate land are held accountable—they are not paid by the Superfund to clean up sites. The Superfund does not regulate closure of sanitary landfills, nor does it promote the conversion of open dumps to sanitary landfills in developing countries.
10. e. A closed sanitary landfill should not be contaminated with hazardous wastes, and thus would not be considered a brownfield.

Chapter 22 – Urbanization and Sustainable Cities

Key Terms

barrio	megacity	slums
brownfield development	megalopolis	smart growth
city	rural area	urban area
conservation development	shantytown	urbanization
greenfield development	squattertown	village

Skills

1. Compare ecological footprints of cities, rural areas, and megacities.
2. Analyze the causes and consequences of urban sprawl.
3. Characterize the problems associated with urbanization in developed and developing countries.
4. Integrate methods that cities can use to be economically, socially, and ecologically sustainable.
5. Describe green urbanism.

> **Take Note:** Understand how the increasing human population is affecting urbanization. The growth of cities has a profound impact on environmental degradation. You must be able to compare the growth of cities in developed and developing countries. Ensure that you are familiar with techniques being used to create more environmentally friendly cities.

Urbanization

Urbanization is an increase in the population in cities. In 2005, 50 percent of the world's population lives in urban areas. Cities have been growing in size since the Industrial Revolution. Countries undergoing demographic transition from an agrarian society to an urban society have a shift in the population as people migrate from rural areas to cities. A rural area is an area where people are dependent upon agriculture or other natural resources for their livelihood. Urban residents do not use natural resources directly as the basis for their occupations. Due to agricultural advancements, fewer farmers can support more people, and therefore fewer people are needed in rural areas. Villages are collections of rural households. A megacity is a city whose population numbers over 10 million. Examples of megacities include Tokyo, Mexico City, São Paulo, New York, Los Angeles, Buenos Aires, Cairo, Delhi, Calcutta, Jakarta, Shanghai, Manila, Osaka, and Beijing. A megalopolis is an area that becomes confluent with urban areas. An example of a megalopolis is the area between Boston and Washington, DC, commonly called Bos-Wash.

Urban growth is due to two main reasons; natural increases in population because the birth rate is greater than the death rate and increases in population due to immigration from rural areas. Natural increase is brought about by sanitation, adequate food supplies, and medical advances that cause the population to increase.

Immigration increases are due to people being drawn into urban areas because of attractive features and because people are displaced from rural areas. Some of the features that attract rural people are jobs, housing, entertainment, and freedom. People are displaced from rural areas for a variety of reasons. Some rural areas are overcrowded and people move to the city to seek employment and housing. Some immigration is due to war, environmental degradation, or racial problems. Poverty has increased in urban areas because rural poor tend to move to cities because conditions for urban poor are better than those of rural poor. Sometimes government policy induces urbanization. Developing countries will spend much more money developing urban areas than they spend on rural areas.

Urbanization in Developing Countries

Large numbers of people result in tremendous amounts of traffic, both vehicle and foot traffic. The traffic, in conjunction with the lack of air quality standards, results in high levels of air pollution in large cities in developing countries. For example, 60 percent of Calcutta's residents suffer from respiratory problems linked to air pollution. Mexico City has one of the highest levels of photochemical smog in the world. Many of these large cities lack adequate sanitation because they do not have modern waste treatment facilities. In Latin America, only 2 percent of urban sewage is treated. One-third of the world population lacks safe drinking water. The waterborne illnesses of cholera, typhoid, dysentery, hepatitis A, and diarrheal diseases are rampant in such areas. In addition to the lack of water, many areas lack adequate housing. One hundred million people in the world are homeless. It is estimated that 20 percent of the world's population lives in inner city slums, shantytowns, or squatter settlements that ring cities. Slums are multifamily dwellings with numerous people per room. Due to the crowding, household accidents such as poisonings and burns are frequently a cause of injury or death to children. Shantytowns (aka *barrios* in South America or *bustees* in India) develop when people build shelters on undeveloped land, and squatter settlements exist where people build on land without the owner's permission. These areas are often located near garbage or industrial waste areas, because these areas were not previously settled.

Urbanization in Developed Countries

Improved sanitation and medical care have decreased the incidence of infectious disease in cities found in developed countries. Due to increased legislation and government policy, air and water pollution have declined in most urban areas. Many older cities are suffering from age-related problems, such as aging sewers, bridges, schools, and streets. Cities are also faced with budgetary issues due to a loss of tax base with so many people moving to the suburbs or to another part of the country. Cities tend to have more poverty, violence, drug abuse, and unemployment than the surrounding suburbs. However, as more people move into suburban areas, these urban problems tend to follow.

In the United States the population has shifted dramatically to the south and west from the north and east, resulting in urban sprawl in these rapidly growing areas. Urban sprawl is the term used to describe dispersed cities that have low population densities. They are the result of inexpensive rural land, availability of automobiles, inexpensive gasoline, and poor urban planning. This population dispersion damages habitat and results in dramatically greater fossil fuel consumption due to the need to commute. In addition, the air pollution problems increase with an increase in suburbs. Two-thirds of all carbon monoxide, one-third of all nitrogen oxides, and one-quarter of all VOCs arise from autos, trucks, and buses. Due to the high concentration of vehicles in many areas, traffic jams are inevitable. Roads are expensive to build, and yet are necessary to accommodate this increased growth in suburban areas.

Smart Growth

Smart growth is the term used to describe the planned development of cities that makes use of existing infrastructure and makes more efficient use of land resources. Smart growth is designed to allow cities and suburbs to grow, while protecting environmental quality. Portland, Oregon, has been very successful in implementing smart-growth programs. Development is restricted to unfilled land within the city and is not allowed to spread to the outer limits of the city, setting up urban growth boundaries. To facilitate this planning, Portland has numerous urban amenities, including an excellent public transit system.

Some states have instituted development rights, in which owners of rural land are paid to prevent development on the outskirts of cities and to preserve the pastoral history of the United States. These programs not only restrict urban growth, they protect farmland for future generations. For example, Maryland's Agricultural Land Preservation Foundation is one of the leaders in farmland preservation. States also may institute a differential tax rate, in which rural areas are taxed at a rate much lower than developed areas, which helps prevent urban sprawl.

Green Cities

Green cities, or green urbanism, is a relatively new concept where cities are redeveloped in a more environmentally sound fashion. These areas may also be called greenfield developments. They have more green space and use more mass transit, ride-share programs, or bicycling routes. These areas may employ reclaimed water systems, have detailed recycling programs, and have green roofs by growing plants on their rooftops to help filter pollutants, absorb water, and cool the area as a microclimate. These cities reclaim brownfields, abandoned industrial sites. The housing is high density and mixed income.

Suburban areas can be designed to preserve the majority of the rural space by clustering the houses instead of spreading them out over a large area. The development cost is less, because electrical, water and sewage, and transportation routes can be clustered as well.

Chapter 22 Questions

1. All of the following cities are megacities except
a. Mumbai b. Los Angeles   c. London d. New York   e. Mexico City

2. The population in the United States has shifted to the
a. Boston, MA   b. Detroit, MI   c. Chicago, ILL   d. Orlando, FL   e. Cincinnati, OH

3. Which of the following has contributed to urban sprawl?
a. increased use of mass transit   b. increased prices for gasoline
c. inexpensive cost of rural land   d. lack of availability of automobiles in urban areas
e. increased jobs in urban areas

4. Which of the following problems do aging cities experience?
a. decreased crime   b. increased tax base
c. decreased poverty   d. increased growth
e. aging infrastructure

5. All of the following contribute to the warmer microclimate in a city except
a. increased particulate pollution causing dust domes.
b. increased transpiration from the abundant vegetation.
c. increased dark surfaces that absorb more sunlight.
d. increased runoff due to impervious surfaces.
e. little shade due to a decline in plant species.

6. Cities in developing countries increase in size due to all of the following except
a. more availability of medical care in cities.   b. political or racial conflicts in rural areas.
c. decreased educational opportunities in cities. d. overcrowding in rural areas.
e. depressed rural employment due to mechanization.

7. Which of the following is characteristic of green cities?
a. increased use of rural space to permit city growth
b. increased vegetation to cool the city
c. reliance on automobiles for transit
d. single-family dwellings
e. increased traffic congestion

8. The proportion of the world's population that reside in inner city slums is thought to approach
____ percent.
a. 5   b. 10   c. 20   d. 30   e. 50

9. Which of the following is not transmitted by contaminated drinking water?
a. typhoid b. dysentery   c. cholera   d. malaria   e. diarrhea

10. Approximately ____ percent of the world's population lacks safe drinking water.
a. 10   b. 30   c. 50   d. 75   e. 90

Chapter 22 Answers

1. c. London is not a megacity, as its population is less than 7 million. Megacities have populations over 10 million.
2. d. The population in the United States has shifted to the south and west, therefore Orlando is more likely the answer.
3. c. The inexpensive cost of rural land has contributed to urban sprawl. Mass transit promotes urban growth, as does increased prices for gasoline. Autos are not a necessity in many urban areas. Jobs would retain people in urban areas.
4. d. Aging cities experience increased crime, lowered tax base, increased poverty, decreased growth, and aging infrastructure.
5. b. Increased transpiration cools vegetated land.
6. c. There are increased educational opportunities in cities in developing countries.
7. b. Characteristics of green cities include decreased use of rural space, increased vegetation to cool the city, fewer automobiles for transit, multiple-family dwellings, and decreased traffic congestion.
8. c. The proportion of the world's population that reside in inner city slums 20 percent.
9. d. Malaria is transmitted by the bite of a mosquito.
10. b. The percentage of the world's population that lacks safe drinking water is approximately 30 percent, or 400 million people.

Chapter 23 - Ecological Economics

Key Terms

classical economics	market equilibrium
cost benefit analysis	neoclassical economics
demand	nonrenewable resource
discount rates	open-access system
ecological economics	political economy
emissions trading	pollution charges
external costs	renewable resource
gross domestic product	resource
gross national product	supply
internal costs	tragedy of the commons
marginal costs and benefits	

Skills
1. Characterize the types of resources.
2. Summarize how supply and demand affect prices.
3. Debate economic growth versus sustainable growth.
4. Review internal and external costs, and explain the purpose of a cost-benefit analysis.

Take Note: Many environmental decisions are made based upon economic policy. A basic understanding of economics is essential in understanding environmental policy decisions. Every AP exam expects students to be able to differentiate between environmental and economic aspects of environmental problems. Ensure that you are familiar with the fundamental concepts of economics in order to be able to adequately address questions that require such understanding.

Economic Terminology

Capital is wealth that can be used to generate more wealth. A resource is anything deemed useful to humans. Resources can be nonrenewable or renewable. A resource is considered nonrenewable if it exists in a finite amount and cannot be replaced within a human lifespan. Nonrenewable resources include metals, fossil fuels, and uranium. Renewable resources have the potential to replace themselves when they are used in a sustainable fashion. Renewable resources include forests, sunlight, and wildlife. Intangible resources are those resources whose value is somewhat abstract, such as finding beauty and peace in a sunset. Classical economics examines how an individual's interests and values mesh with societal goals. Modern-day political economists still study moral philosophy and social structure. The neoclassical economists use science processes to study economic analysis. Neoclassical economists claim to be free of judgments and values when addressing economic issues.

Supply and Demand

Demand is the amount of a product or service consumers are willing to buy at various prices. Supply is the quantity being offered for sale. Classical economics believe there is an inverse correlation between supply and demand. As the price for a good rises, the supply increases and the demand declines. Supply and demand meet at an intersection called the market equilibrium. However, prices are not controlled just by supply and demand. Marginal costs and benefits, or the additional cost of more units of the good or service, also plays a role in the cost of an item. Price inelasticity exists when a rising price does not decrease demand. Some items follow supply and demand curves exactly and are said to have price elasticity.

Ecological Economics

Ecological economics takes into account impacts on the environment and attempts to include natural services into the price of a resource. Internal costs are the costs directly related to obtaining a resource. The internal costs of an automobile would be the expense of obtaining the materials to build the car, the cost of the automobile workers and sales people, the costs of advertising the car, and so on. External costs are costs such as pollution, road damage, or heath care costs due to emissions from the automobile. If you incorporate the internal costs with external costs, it is called full-cost pricing. The full-cost pricing of the automobile in our example would reflect the environmental damage associated with using the car. Such costs would likely be prohibitively expensive.

Tragedy of the Commons

In 1968 Garrett Hardin wrote an article entitled "Tragedy of the Commons." The article explains that any common resource tends to be over used because all people using the resource will use as much as they can without consideration for the greater good of the rest of the population. Examples would be whaling, air pollution, and the decline of fisheries throughout the world. An open-access system is one in which there are no limits to resource use. Many societies throughout time have found solutions to open-access systems, usually by expecting cooperation between users.

Scarcity

Many times scarcity promotes unique solutions to seemingly impossible problems. Increased technological advances may shift supply and demand curves. Greater production may lead to greater availability, thus dropping prices. When materials become scarce, other materials may need to be substituted for the original. For example, aluminum wiring in electrical components became common when copper prices increased. Recycling also increases when supplies are limited.

Human Carrying Capacity

Humans are the only organisms on the planet with the intelligence to increase their carrying capacity by employing technology. As we increase the numbers of people on the planet, most ecologists feel that we will eventually reach a point where the environment can no longer support such large numbers of people. Most economists believe that technological advances and substitution for scarce items will allow human populations to continue to increase.

Gross National Product

The gross national product (GNP) is the cost of production of all of the goods and materials produced by a country in a given year. For example, all of the automobiles made by a U.S. company and sold throughout the world are calculated in the GNP. The gross domestic product (GDP) is the cost of production of all of the goods and services produced for use by a country in a given year, and therefore includes only the economic activity that takes place within the country itself. For example, chemicals made and sold by a U.S. company in another nation would not be included in the GDP. These values are somewhat misleading, because they do not reflect quality of life, disease rates, destruction of natural resources, and other indicators of environmental damage. The genuine progress index (GPI) does take into account quality of life, depletion of natural resources, and environmental damage. The human development index (HDI) evaluates life expectancy, educational opportunities, and standard of living as measurements involved in development. When the HDI is adjusted for gender, the gender development index (GDI) is created. Development levels are high in Europe, North America, and Japan and low in Africa and Haiti.

Cost Benefit Analysis

The cost benefit analysis (CBA) assigns values to resources and their recovery, including the environmental damage incurred. The goal is to determine the amount of pollution control that yields benefits for the lowest cost. For example, a CBA on logging would take into account money spent on land, materials, and labor, which are fairly straight forward. The other intangible values lost, such as aesthetic value, relaxation, and solitude, are difficult to place a monetary value upon. It is difficult to place these value measurements into a formula to generate a cost benefit analysis.

Economy Based Pollution Reduction

When cleaning up pollutants, the price of cleanup goes up the fewer units of the pollutant there are in the environment. A good analogy is spilling a box of packing peanuts. The large pile formed is easy to scoop back in the box, but it takes more time and energy (money) to pick up the individual peanuts that have scattered to the corners of the room. The polluter may determine that there is a marginal cost of pollution abatement, or an economically feasible amount of pollution to release based upon cleanup costs or pollution fees. Pollution charges, green taxes, or effluent fees are used to charge or tax polluters per unit of effluent released. More money is saved if pollution is reduced. Emissions trading involves a market-based system that allows polluters to sell or trade emissions credits if they release below the level of pollution they are allowed to emit by law. The polluting company can make money selling or trading its credits only as long as it is below the levels permitted. Therefore, it is beneficial to the company to emit less pollution so that they may sell their credits. In the United States the Clean Air Act allows sulfur dioxide pollution credits. Under the Kyoto Protocol, carbon dioxide credits are being traded internationally. Discount rates are basically interest rates applied to future benefits. Forecasting future benefits is difficult, and thus this method of accounting for environmental damage has many opponents.

Global Trade

There are two major international agreements that govern most international trade, the General Agreement on Tariffs and Trade (GATT) and World Trade Organization (WTO) agreements. The WTO handles trade disputes, administers trade agreements and negotiations, and assists developing countries with training and technical assistance. Critics of these agreements cite the environmental damage done to developing countries as a result of trade that depletes their natural resources to better supply developed countries. The North American Free Trade Agreement (NAFTA) is likewise opposed because American companies will move to Mexico to exploit their more lax laws regarding labor, environmental degradation, and pollution. The World Bank was designed to assist with aid for Europe and Japan after World War II. It later became involved in assisting developing countries with development. The loans provided by the World Bank have given developed countries access to markets and resources previously unavailable. One of the problems associated with the loans is the environmental damage that can be incurred by development. Microlending is small-scale lending to promote development. It is designed to provide families with a means of support, such as a loom, cow, or sewing machine. These loan programs are successful, because the amounts are small and the repayment rates are high.

Green Business

Many companies have determined that pollution prevention is much less expensive than pollution cleanup. These companies recycle within the company (preconsumer), purify their wastewater, use less hazardous materials during production, and lower their emissions and solid waste production. Companies dedicated to these programs include 3M, DuPont, and Monsanto. These companies report their value with a triple bottom line, which takes into account environmental effects and social justice as a component of their success. Consumers can increase the success of green business by purchasing green products, known as green consumerism. Green business does not mean fewer jobs. On the contrary, jobs are created by new, innovative methods to reduce waste, to reuse products, and in recycling.

Questions Chapter 23

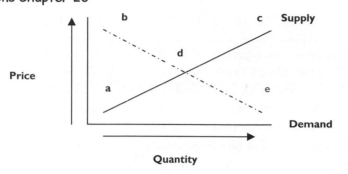

1. In the diagram above, which letter denotes market equilibrium?

2. If the price of petroleum doubles, which of the following statements would be correct?
a. The supply will not change, because all of the known reserves are being used.
b. The supply will increase as new reserves are tapped due to the increased economic feasibility of recovering the petroleum.
c. The supply will decline dramatically, because petroleum will be too expensive to recover.
d. The supply will decline. Because petroleum is so expensive, no one can afford to purchase items produced from the refining of petroleum.
e. The supply will increase as more petroleum is produced naturally.

3. All of the following are renewable resources except
a. oxygen.   b. forests.   c. natural gas.   d. topsoil.   e. deer.

4. All of the following are examples of intangible resources except
a. solitude.   b. beauty.   c. relaxation.   d. timber.   e. satisfaction.

5. Which of the following is the best example of tragedy of the commons?
a. a group of hunters harvesting deer according to established hunting regulations
b. long-line fishermen killing numerous nontarget species in their by-catch
c. a logging company using shelterwood cutting for a forest
d. a company releasing less CO_2 and SO_2 from its emissions than required by law
e. a rancher building retention ponds on his land around pastures to decrease runoff into surrounding streams

6. Which of the following is not taken into account when calculating a country's GNP?
a. production of textiles in the southeastern United States
b. wireless Internet installation
c. mining of coal in West Virginia
d. the infant mortality rate in Illinois
e. income made by a computer company

7. An external cost of owning and operating an automobile would be
a. cost of the automobile.   b. cost of the gasoline.   c. cost of the tires.
d. cost of registering the vehicles.   e. pollution emitted by the automobile when driven.

8. Which of the following is an example of a green tax?
a. allowing an individual to receive a $500 tax deduction for purchasing an electric car
b. charging a coal burning power plant a tax for any SO_2 released over the level permitted by law
c. subsidizing companies that specialize in making photovoltaic cells
d. charging a lower sales tax on hybrid cars
e. a city giving a rebate for purchase of a low-flow toilet

9. Tradable pollution permits for sulfur dioxide are advantageous because
a. they reduce pollution at the plants that pollute the most.
b. they encourage companies to prevent pollution.
c. small companies may not be able to purchase or trade the permits.
d. the pollution caps may be set too low allowing more acid deposition.
e. companies that pollute the most do not have to decrease their emissions.

10. Which of the following organizations regulates international trade?
a. World Bank   b. NAFTA   c. WTO   d. GDP   e. HDI

Chapter 23 Answers
1. d. In the diagram, the letter d denotes market equilibrium as it is the junction between supply and demand.
2. b. The supply will increase as new reserves are tapped due to the increased economic feasibility of recovering the petroleum. Not all of the known reserves are being used due to economic feasibility or technological difficulty. The price increasing will cause the demand to decline. The supply cannot naturally increase in a feasible amount of time.
3. c. Natural gas is a fossil fuel, and therefore is a nonrenewable resource.
4. d. Timber is a tangible resource.
5. b. Long-line fishing and the large numbers of nontarget animals killed as by-catch are an example of tragedy of the commons.
6. d. The infant mortality rate is not taken into account when calculating a country's GNP.
7. e. The pollution emitted by the car would be an external cost of owning and operating an automobile.
8. b. Green taxes, a "stick," are taxes on effluent. They are different from tax deductions, subsidies, and rebates, which usually are "carrots."
9. b. Tradable pollution permits for sulfur dioxide are advantageous because they encourage pollution prevention. The other aspects are drawbacks to tradable permits.
10. c. The World Trade Organization, WTO, regulates international trade.

Chapter 24 - Environmental Policy, Law, and Planning

Key Terms

administrative courts	environmental law	precautionary principle
administrative law	environmental policy	precedent
arbitration	globalization	rider
case law	green plans	statute law
civil law	lobbying	tort law
common law	mediation	
criminal law	policy	

Skills

1. Describe how policies are established.
2. Summarize the steps required to make a federal law.
3. Contrast the differences between civil, criminal, and administrative laws.
4. Assess the effectiveness of global treaties.

> **Take Note:** The AP Environmental exam frequently has an essay that asks you to recommend a policy and then give reasons why or why not the policy will be successful. This policy has to be feasible, but can be a policy promoted by incentives or disincentives. For example a policy that may promote the use of hybrid automobiles would be to increase the tax on gasoline. A policy that would encourage a consumer to switch to a low-flow shower head might be a rebate from a utility company toward the purchase of the shower head.

Environmental Policy

Policy is a plan about how to accomplish a goal. Environmental policies are developed to implement and enforce regulations regarding environmental issues. Policies are often created by interest groups pursuing their own goals. Utilitarian benefits must be considered when making decisions regarding environmental issues, but the aesthetic, cultural, intrinsic, and other intangible benefits must also be taken into account. When a policy is being devised, proposals are developed to solve a problem. Laws are enacted to implement the policy and their efficacy is evaluated over time.

Environmental Law

Environmental law refers to the legislation used to protect environmental quality, natural resources, and sustainability. Statute laws arise from legislative bodies. Case law involves application of the statute laws as decided by court decisions. Administrative law results from executive orders and administrative policies.

One of the most important pieces of legislation is the National Environmental Policy Act (NEPA) of 1969. It established policies to protect the environment from humans by requiring environmental impact statements for construction projects regulated by federal government.

Statutory laws are passed after a bill is ratified by both houses of Congress and signed by the president. A bill is subject to a series of hearings to determine its scope. Once the bill has been through the hearings and has support, it is passed from the appropriate subcommittee to the main floor of the legislature, where it is further debated. Once each house has its final version, representatives and senators meet to finalize the text. If the president vetoes the bill, the houses can override his veto with a two-thirds majority vote. The president may also elect not to sign it instead of outright vetoing it. In that event, it becomes law after ten days.

Legislators often attach riders to a bill to pass unrelated or controversial laws. Appropriations bills are often the bills to which these riders are attached. For example, if Congress is passing funding for necessary hurricane relief, a rancher in Montana may have his or her legislator attach a rider permitting damage to a wild and scenic river. The funding for hurricane relief must be passed quickly, and thus the rider passes, as well. Lobbyists are individuals that promote the agenda of a particular group to the Congress. Media attention is often used by lobbyists to promote their individual agendas.

The judicial branch enforces the laws passed by the legislature. Their interpretation of the law creates case law. Precedent is established when the first judge to hear a case interprets the law in a particular way and makes a judgment. These decisions are only applicable to lower courts. If a district court decision moves to the appellate court and wins, the case goes to one of 12 regional federal courts. The Supreme Court decides to hear a certain number of cases appealed above the level of the federal court, and they are the final step in determining the interpretation of a law.

Criminal law involves state or federal law that prohibits wrongdoing. Serious crimes are felonies, and those of lesser severity are misdemeanors. A criminal prosecutor takes a defendant to court, which determines his or her guilt or innocence.

Civil law is the body of laws that regulate individuals and corporations. Civil cases can be decided on a preponderance of evidence. Civil law cases in environmental law often involve an individual or group of individuals (class) suing another group for environmental damage. There is no jail time in civil cases, just financial settlements. Common laws are laws that have been established by historical precedent, but not passed by a legislative body. Tort law is law in which one party seeks damages against another party.

Administrative law is the establishment and monitoring of rules set up by government administrators. For example, the Environmental Protection Agency (EPA) can set energy-efficiency standards on home appliances. Executive orders, or statements of law made by the president, are also examples of administrative law. In 1993 President Clinton signed an executive order requiring the federal government to purchase paper that contained at least 30 percent postconsumer recycled paper. As one of the world's biggest users of paper, the federal government generated a demand for recycled paper by instituting this order.

The federal government is broken into a series of agencies responsible for administering laws. The EPA regulates air, water, and noise pollution; radiation, pesticide, and toxin regulation; and solid waste disposal. The Department of the Interior has several different departments and offices. The Office of Surface Mining regulates surface mining and reclamation of coal deposits. The National Park Service manages U.S. national parks, and the Bureau of Land Management manages public lands in western United States. The U.S. Fish and Wildlife Service carries out ecosystem management and endangered species plans while the Bureau of Reclamation is responsible for water management in western United States. The U.S. Geological Survey reviews natural resources, landscape, and natural hazards, and the Minerals Management Service manages fossil fuels and minerals on the outer continental shelf. The Department of Agriculture contains the U.S. Forestry Service and the Natural Resources Conservation Service. The Department of Energy regulates fossil fuels, nuclear energy, renewable energy resources, and nuclear wastes.

Dispute Resolution

Rather than becoming embroiled in a lengthy court battle over issues, many individuals and companies subject their disagreement to arbitration, where an unbiased third party makes a binding decision regarding the issue in question. Arbitration is not constrained by precedent; otherwise the procedure is similar to a trial. Mediation is when a third party comes into a disagreement to help two parties find some accord. The job of a mediator is to facilitate communication, but they do not make a final decision. Both methods can be less expensive and faster than going to court, and thus are favored in some situations.

International Treaties and Agreements

Many of the international agreements on the environment are just that—agreements. They are nonbinding and unenforceable agreements regarding environmental concerns. The United Nations (UN) may become involved as needed, but countries that do not sign the agreements or countries that sign it and then violate it are not typically subject to any penalty if they violate its rules. Environmental groups will often attack national policies that violate these accords, and the ensuing embarrassment is many times the impetus to affect change in those nations. Globalization, the increasing amount of communication and commerce between nations, is increasing the impact of environmental decisions.

Precautionary Principle

The precautionary principle states that precautionary measures should always be taken if it is suspected that an activity could cause harm to humans or the environment. The European Union (EU) has adopted the precautionary principle as its basis of environmental policy, but the United States feels that it will impede economic advances. American companies that have European branches must adhere to EU rules regarding the environment.

Green Plans

Green plans are long-term environmental strategies that have been established by some countries, including Canada, Sweden, and Denmark. These plans reduce pollution while increasing economic stability.

Chapter 24 Questions

Use the following for questions 1-4.

    a. NEPA
    b. CITES
    c. CERCLA
    d. EPA
    e. ESA

1. regulates international trade in endangered and threatened species
2. requires that an environmental impact statement is written for any federal project
3. agency established to enforce environmental laws
4. law allowing rapid response and cleanup of hazardous waste sites

Use the following for questions 5-8.

    a. tort law
    b. case law
    c. administrative law
    d. statute law
    e. common law

5. established by a legislative body
6. established by court decisions
7. involves monetary settlements
8. established by the executive branch and its agencies

9. Riders
a. must be related to the bill to which they are attached.
b. are often controversial but passed due to the importance of the original law.
c. are added by Congress after a bill has been signed into law.
d. are added by the president as he signs a bill into law.
e. are added by the courts to clarify laws.

10. The international agreement that called for a reduction of CFCs is
a. CITES.   b. NEPA.   c. The Basel Convention.
d. the Montreal Protocol.   e. the Kyoto Protocol.

Chapter 24 Answers

1. b. CITES regulates international trade in endangered and threatened species.
2. a. NEPA requires that an environmental impact statement is written for any federal project.
3. d. The EPA was established to enforce environmental laws.
4. c. CERCLA is the law allowing rapid response and cleanup of hazardous waste sites.
5. d. Statute law is established by a legislative body.
6. b. Case law is established by court decisions.
7. a. Tort law involves monetary settlements.
8. c. Administrative law is established by the executive branch and its agencies.
9. b. Riders are often controversial but passed due to the importance of the original law.
10. d. The Montreal Protocol is the international agreement that called for a reduction of CFCs.

Chapter 25 – What Then Shall We Do?

Key Terms

deep ecology
environmental literacy
green political parties

nongovernmental organizations
sustainable development
wise use groups

Skills

1. Identify the difference between radical and mainstream environmental organizations.
2. Characterize green political party involvement in sustainable development.
3. Compare the wise use movement to deep ecology.

Take Note: This last chapter of the text explains what an individual can do to make a difference in the issues discussed in the text. Little of the information in the chapter will be of relevance on the AP exam. One of the most important concepts to take away from this chapter is that individuals *can* make a difference, by shutting off lights in a room you are not in or by shutting off the water as you brush your teeth. By reading this textbook and enrolling in this class, you have been better educated in the problems facing humans on this planet. When asked on an AP exam to recommend policy for environmental change, education will nearly always be accepted as an answer, when properly explained. Educating the public is the first step in making people aware of the changes that need to be made to create a sustainable environment.

Environmental Education

The federal government deemed environmental education to be important and passed the National Environmental Education Act to help teach people about sustainability and global environmental issues and to encourage students to pursue careers in environmental science. Environmental literacy is the ability of all people to understand basic environmental terminology and concepts.

Individual Involvement in Environmental Issues

As an individual, you can decide to decrease consumption by reducing your ecological footprint. You can purchase less "stuff," you can avoid disposable items and excess packaging, and you can conserve water and save energy. You can buy products from green businesses that help maintain a sustainable environment. You can ensure that products you buy have little impact, such as those labeled "green," but be a wary consumer, because some items labeled as "green" might be falsely identified as environmentally safe.

For example, when you grocery shop and the clerk says "paper or plastic" many people pick paper because they know that paper is derived from a renewable resource and plastic is derived from oil. In fact, paper production is far more polluting of both air and water than is plastic production. Grocery bags have to be made from virgin wood and cannot be recycled into grocery bags again due to the fiber length. Therefore they must enter an open-loop recycle plan and cannot be reused as paper bags. Paper bags are also more expensive to ship than the lighter plastic bags, and plastic takes up less landfill space than paper does. Therefore, neither choice is as good as bringing your own cloth bags from home.

There are many ways to influence environmental policy. You can by voting, writing your legislators, and joining an environmental organization. A new type of politics, called Green Parties, has emerged as interest in the environment has picked up. The most powerful party is in Germany. In the United States, the strong two-party political system holds a third party at bay, but green politicians have been elected to some local and state offices. Some of the mainstream environmental organizations are the National Wildlife Federation, World Wildlife Fund, Sierra Club, Audubon Society, and the Wilderness Society. Some groups are radical, like Earth First! and Sea Shepherd. They follow the tenets of deep ecology, the belief that humans must become more environmentally oriented. Some groups have environmentally friendly sounding names, such as the National Wetlands Coalition, but are a group of companies that include mining and oil companies that want to weaken wetlands protection. They are part of the wise use movement, whose goal is to weaken all environmental legislation, primarily in the interest of economics.

Global Action

All humans must learn to live sustainably. Sustainable development is the improvement of the standard of living for individuals without harming the environment or using potentially renewable resources at a rate at which they cannot replenish. Some of the goals of sustainable development include low birth and death rates via demographic transition, energy use that is high efficiency and depends on potentially renewable resources, and economic transition toward sustainability. Nongovernmental organizations, or NGOs, are groups that work for social change. Two examples are the Stockholm Conference of 1972 and the Rio Earth Summit of 1992. Some of the NGOs work as private organizations, but others assist the governments of countries to promote change.

The UN Conference on Environment and Development in Rio de Janeiro in 1992, aka Rio Earth Summit, created Agenda 21, a list of sustainability goals. Topics included biodiversity, deforestation, global climate change, rights of indigenous people, poverty, and sustainable development. A UN council created the Earth Charter shortly thereafter, which listed principles of environmental protection. The following table lists the principles of the Earth Charter.

Table 25.1 Earth Charter Principles

1. Respect Earth and life in all its diversity.
2. Care for the community of life with understanding, compassion, and love.
3. Build democratic societies that are just, participatory, sustainable, and peaceful.
4. Secure Earth's bounty and beauty for present and future generations.
5. Protect and restore the integrity of Earth's ecological systems, with special concern for biological diversity and the natural processes that sustain life.
6. Prevent harm as the best method of environmental protection and, when knowledge is limited, apply a precautionary approach.
7. Adopt patterns of production, consumption, and reproduction that safeguard Earth's regenerative capacities, human rights, and community well-being.
8. Advance the study of ecological sustainability and promote the open exchange and wide application of the knowledge acquired.
9. Eradicate poverty as an ethical, social, and environmental imperative.
10. Ensure that economic activities and institutions at all levels promote human development in an equitable and sustainable manner.
11. Affirm gender equality and equity as prerequisites to sustainable development and ensure universal access to education, health care, and economic opportunity.
12. Uphold the right of all, without discrimination, to a natural and social environment supportive of human dignity, bodily health, and spiritual well-being, with special attention to the rights of indigenous peoples and minorities.
13. Strengthen democratic institutions at all levels, and provide transparency and accountability in governance, inclusive participation in decision making, and access to justice.
14. Integrate into formal education and life-long learning the knowledge, values, and skills needed for a sustainable way of life.
15. Treat all living beings with respect and consideration.
16. Promote a culture of tolerance, nonviolence, and peace.

Source: The Earth Council, San Jose, Costa Rica, 2001.

Chapter 25 Questions

1. Which of the following would the wise use movement support?
a. creating a preserve on state land to protect threatened species of rose orchids
b. filling in a wetland to promote development of urban areas
c. replanting a deforested area in South America that had been logged for rangeland
d. setting aside remaining tall grass prairie in Iowa to preserve the biome
e. preventing drilling in ANWAR to allow caribou calving grounds to be undisturbed

2. Which of the following would not be classified as an NGO and its subject?
a. the environment—The Earth Council
b. women's rights—Center for Reproductive Rights
c. peace—Good Neighbors International
d. human rights—Doctors Without Borders
e. hunting and trapping wildlife—Abundant Wildlife Society of North America

3. Which of the following is a radical environmental group?
a. Sierra Club   b. Nature Conservancy   c. Natural Resources Defence Council
d. Earth First!   e. World Wildlife Fund

4. Which of the following is not an example of green shopping?
a. buying disposable razors
b. purchasing organically grown strawberries
c. purchasing rechargeable nickel cadmium batteries
d. buying notebook paper composed of 30 percent postconsumer recycled paper
e. buying two liter bottles of soft drinks instead of six packs of cans

5. All of the following are part of the Earth Charter principles except
a. developing ecological sustainability.
b. strengthening democratic institutions and societies.
c. developing economic strategies without regard to social welfare.
d. treating all living beings with respect.
e. preventing harm to the environment.

Chapter 25 Answers

1. b. Filling in a wetland to promote development of urban areas would be approved of by the wise use movement. The other examples protect habitat and species.
2. e. All are NGOs with their correct subject except e. The Abundant Wildlife Society of North America promotes hunting and trapping of wildlife. The group formed in response to the plan to repopulate Yellowstone with wolves. NGOs are large influential organizations that work for social change.
3. d. Earth First! is a radical environmental group. All of the others are mainstream groups.
4. a. Buying disposable razors is not an example of green shopping. You should purchase razors where just the blades can be replaced or use a rechargeable electric razor.
5. c. The Earth Charter specifically states that economic activities and institutions promote human development in an equitable and sustainable manner.

- 228 -

Appendix A
Important Environmental Legislation (in chronological order)

Mining Act of 1872 – allows companies to purchase federal land if they patent it

1897 – Forest Management Act - creates the Forest Reserves to be used for timber, mining, and grazing.

Lacey Act of 1900 - prohibits the transport of live or dead wild animals or parts of wild animals across state lines without a federal permit

Antiquities Act of 1906 – establishes areas of archeological/or historical interest on federal lands as monuments

National Park Service (NPS) Act of 1916 – establishes the NPS

Migratory Bird Treaty Act of 1918 – places hunting restrictions on migratory birds

Taylor Grazing Act of 1934 – allows Department of the Interior to issue grazing permits and collect fees to graze animals on public land

Soil Conservation Act of 1935 – establishes the Soil Conservation Service in the United States Department of Agriculture (USDA); focused on erosion control

Federal Aid in Wildlife Restoration Act of 1937 – aka Pittman-Robertson Act – provides federal funds for wildlife protection

Federal Food, Drug and Cosmetic Act of 1938 – protects consumers

Federal Insecticide, Fungicide, and Rodenticide Act of 1947 (FIFRA) - protects from dangerous and persistent pesticides

Price-Anderson Act of 1957 – promotes nuclear power by limiting the liability of power plant owners or the government in the event of a major accident

Delaney Clause – 1958 – amendment to the Federal Food, Drug and Cosmetic Act of 1938 prohibits the addition of any known carcinogen to any processed food, drug, or cosmetics

Multiple Use and Sustained Yield Act of 1960 – national forests are to be used for recreation and hunting, not just timber and mining

Clean Air Act of 1963 – air quality standards

Wilderness Act of 1964 – established the National Wilderness Preservation System which protects primitive areas in the United States

Federal Water Pollution Control Act of 1964 – aka **Clean Water Act** – restore and maintain U.S. waters; amended in 1972 and 1977 to regulate discharges of pollutants into water; gives EPA power to implement pollution control programs; sets water quality standards in surface waters; regulates point source pollutants; funds the construction of sewage treatment.

National Wild and Scenic Rivers Act of 1968 – protects unique and beautiful river sections

National Environmental Policy Act of 1969 (NEPA) – requires environmental impact statements for construction projects regulated by federal government

Clean Air Act of 1970 – protects air as a resource from pollution; the EPA is to establish National Ambient Air Quality Standards (NAAQS) to protect public health and the environment; amended in 1977 to set new dates for reaching NAAQS; 1990 amendments to the Clean Air Act addressed acid rain, ground-level ozone and stratospheric ozone depletion

FIFRA of 1972 – registration and testing of pesticides required

Marine Protection, Research, and Sanctuaries Act of 1972 – establishes the National Marine Sanctuary Program administered by the National Oceanic and Atmospheric Administration

Ocean Dumping Act of 1972 – prohibits dumping of sewage sludge and industrial waste to include radiological, chemical, or biological warfare agents, high-level radioactive waste, and medical waste

Marine Mammal Protection Act of 1972 – protects marine mammals

Endangered Species Act of 1973 – USFWS and USNMF must establish programs to protect endangered organisms

Safe Drinking Water Act of 1974 – sets standards and regulates public drinking water

Energy Policy and Conservation Act of 1975 – promotes conservation and efficiency

Federal Land Policy and Management Act of 1976 – Bureau of Land Management (BLM) land is managed under multiple-use, sustained-yield principles

Toxic Substances Control Act of 1976 – regulation of toxins

Resource Conservation and Recovery Act of 1976 - (RCRA) – tracking of hazardous wastes; states develop hazardous waste management plans

Surface Mining Control and Reclamation Act of 1977 - (SMCRA) - regulates restoration of surface mines

Comprehensive Environmental Response, Compensation and Liability Act of 1980 - (CERCLA or Superfund) – uses Superfund to clean up hazardous waste sites (new regulations established with SARA – Superfund Amendments and Reauthorization Act in 1986)

International Treaties to Know

CITES – Convention on International Trade in Endangered Species of Wild Fauna and Flora – 1987 – Regulates international trade in endangered species

Montreal Protocol on Protecting Stratospheric Ozone - 1987 – phases out use and production of CFCs

Basel Convention on the Transboundary Movements of Hazardous Wastes and Their Disposal – (1992) prevents disposal in developing countries of hazardous wastes produced in developed countries

Convention of the Prevention of Marine Pollution by Dumping of Waste and Other Matter (London Dumping Convention or LDC) – controls ocean pollution; prohibits dumping of high-level radioactive wastes, heavy metals, and other hazardous wastes

Important Government Offices

Environmental Protection Agency – regulates air, water, and noise pollution; radiation, pesticide, and toxin regulation; solid wastes

Department of the Interior
    Office of Surface Mining—regulates surface mining and reclamation of coal deposits
    National Park Service—manages U.S. national parks
    Bureau of Land Management—manage public lands in western United States
    U.S. Fish and Wildlife Service—ecosystem management and endangered species plans
    Bureau of Reclamation—water management in western United States
    U.S. Geological Survey—reviews natural resources, landscape, and natural hazards
    Minerals Management Service—manages fossil fuels and minerals on outer continental shelf

Department of Energy – regulates fossil fuels, nuclear energy, renewable energy resources, and nuclear wastes

Department of Agriculture – contains the U.S. Forestry Service and the Natural Resources Conservation Service

Appendix B
Major Pollutants Chart

Pollutant	Located on	Source(s)	Environmental Impact	Human Health Effects	Prevent/Remediate
Aluminum	Soil; naturally occurring element	Naturally occurring	More soluble if acidic conditions Toxic to fish unless pH > 5.5; High levels of Al toxic to plants; Does not bioaccumulate	Pulmonary problems The FDA has ruled Al in cooking utensils, foil, antacids, and antiperspirants is safe	Prevent—maintain pH in ecosystems by preventing acid deposition Remediate—increase the pH with buffers to decrease solubility
Arsenic	Soil; water; naturally occurring element	Used in wood preservation; chemical processes; petroleum mining, mining and smelting Was formerly used in pesticides, rodenticides, and herbicides Humans usually ingest arsenic in food or drinking water	Plant toxicity—wilt, brown, die Animal toxicity—causes aquatic organisms to have decreased growth; metabolic failure in many species	Acute—anemia; nausea Chronic—carcinogen, teratogen likely mutagen, induces chronic fatigue, gastrointestinal disease, anemia, death Acceptable levels in U.S. drinking water is 10 ppb	Prevent—remove arsenic from drinking water; government regulatory action; use other wood preservatives Remediate—phytoremediation with brake fern, which readily removes arsenic; some bacteria can oxidize arsenic
Asbestos	Air (indoor); natural mineral fiber	Added to materials for strengthening and for fire resistance in ceiling tiles, floor tiles, roofing, shingles and siding, textured paints and joint compounds, brake pads, and firefighter equipment	Human health effects	Chronic—mesothelioma, lung cancer, asbestosis (scarred lungs from inhaling fibers)	Prevent—banned use in paint/joint compound in 1977; government regulatory action; cover the asbestos with another material; use alternative materials as fire retardants Remediation—removal by qualified professionals
Benzene	Air water; naturally in petroleum; produced by forest fires and volcanoes	Used as a solvent to make plastics, rubber, and synthetic textiles; gasoline combustion; petroleum refining; ETS	Toxic to aquatic species; Kills plants by damaging roots and leaves Does not bioaccumulate	Acute—anemia, depressed nervous and immune function Chronic—cancer, gene mutation, known human carcinogen	Prevent—avoid spills Remediate—can be broken down by bacteria
Cadmium	Soil; Naturally occurring element	Electroplating; smelting; chemical processes; incineration	Toxic to plants Toxic to wildlife; bioaccumulates in liver and kidney	Acute—affects respiratory system, muscle contractions Chronic—Carcinogen, teratogen, affects growth; reproduction; kidney disease; hypertension	Prevent—Government regulatory action; use an alternative metal in chemical processes Remediate—Phytoremediation with pennycress; bioremediation with bacteria

Pollutant	Location	Source(s)	Environmental Impact	Human Health Effects	Prevent/Remediate
CFCs	Air—move from troposphere to stratosphere	Used as solvents; propellants; refrigerants; foam blowing agents Includes methyl chloroform (solvent), halons (fire extinguishers), methyl bromide (crop fumigant), carbon tetrachloride (solvent) and freons (refrigerant)	Chlorine radicals convert ozone to oxygen gas $2O_3 \rightarrow 3O_2$ Increased UV radiation pass through stratosphere Damage to living organisms due to increased UV CFCs are greenhouse gases	Health effects indirect; due to increased ultraviolet radiation Increased skin cancer; cataracts; immunosuppression	Prevent—use other refrigerants and propellants; government regulatory action Difficult to remediate
CO	Air	Incomplete combustion of fossil fuels; gas furnace/space heaters; gas/wood stoves; ETS; car exhaust	Minimal environmental impact except human health; no direct effect on plants	Blood's ability to carry oxygen is impaired Acute—fatigue, impaired vision, headache, dizziness, nausea, confusion High concentrations are fatal	Prevent—maintain wood/gas appliances; catalytic converter in autos; decrease use fossil fuels; government regulatory action Remediate—increase ventilation; improve efficiency
CO_2	Air	Combustion of biomass, solid waste, or fossil fuels		Very little unless CO_2 increases dramatically as in a confined space; then causes acidosis Climate change due to CO_2 will have more human health impacts	Prevent—government regulatory action; use alternative energy resources; maintain forests; improve energy efficiency to decrease fossil fuel consumption; mass transit, carpooling, bicycles to decrease CO_2 emissions from vehicles Remediate - Plant vegetation; Pump carbon dioxide into underground storage
Copper	Soil; water	$CuSO_4$ (copper sulfate)—Algicide	Plant micronutrient Highly toxic to amphibians; bioaccumulates in some species	Acute—nausea, vomiting, cough, headache, difficulty breathing Chronic—liver cirrhosis; low blood pressure; fetal mortality; kidney and brain necrosis	Prevent—use mechanical or biological methods to remove algae Remediate—electric current can remove copper from contaminated soil

Pollutant	Location	Source(s)	Environmental Impact	Human Health Effects	Prevent/Remediate
Cyanide	Soil; water	HCN; CN⁻ Gold mining; industrial discharge	Fish very sensitive—does not bioaccumulate; not persistent Sublethal fish effects—decreased reproduction; growth altered	Acute—suppresses aerobic respiration; rapid breathing and tremors Chronic—weight loss, nerve damage	Prevent—avoid release; utilize other chemical processes Remediate - bioremediation; Chemical remediation
Dioxins	Soil; water	Group of similar chemical compounds Not produced intentionally; naturally produced in forest fires; anthropogenic sources include incineration; chlorine bleaching at paper mills; ETS Most toxic is TCDD Human exposure in fish ingested from contaminated waterways; ingestion of animal fat	Bioaccumulates in fat Animal toxicity—Liver toxicity; affects endocrine, immune, nervous and reproductive systems Highly persistent	Acute—chloracne, skin rashes, skin discoloration Bioaccumulates in fat Chronic—liver damage, teratogen, immunotoxic, likely human carcinogen	Prevent—government regulatory action, use chemicals other than chlorine to bleach or sterilize, remove plastics containing chlorine prior to waste incineration, pollution control devices to remove dioxins after incineration Remediate - bioremediation; Chemical remediation; Physical removal of contaminated soil
Disease agents	Water	Animal and human wastes Indicated by increased levels of fecal coliforms, which may indicate presence of human pathogens	Includes bacteria that cause cholera, typhoid, , dysentery, and viruses like hepatitis A May also include *Cryptosporidium* and *Giardia*, protozoal parasites that cause gastrointestinal disease	Causes human disease including diarrhea, nausea, and so on.	Prevent—ensure wastes are treated prior to discharge into surface water or groundwater, chlorination, government regulatory action Remediate - treat water to kill organisms
Environmental Tobacco Smoke (ETS)	Air (indoor)	Cigarette smoking	Human health effects	Lung cancer, respiratory disease, heart disease	Prevent—improve ventilation, prevent smoking indoors, government regulatory action Remediate - ventilate; Clean area thoroughly
Formaldehyde	Air (indoor)	Released from building material such as plywood, textiles, furniture stuffing, carpets	Human health effects	Acute/chronic—dizziness, rash, breathing problems, headaches, and nausea	Prevent—use other materials to manufacture materials, use other materials that do not contain formaldehyde Remediate—improve ventilation

Pollutant	Location	Source(s)	Environmental Impact	Human Health Effects	Prevent/Remediate
Lead	Air, soil, water	leaded gasoline, leaded paint, leaded solder, lead shot/sinkers, contaminated soil, smelters, incinerators, utilities, automobile batteries	Plants—decreased growth; decreased photosynthesis Animals—CNS damage; sterility; effects similar to human exposures	Acute—abdominal pain, fatigue, sleep disturbance, high blood pressure; death Bioaccumulates in bone Chronic—developmental retardation, impaired IQ, attention deficit, hyperactivity, learning disorders, aggression, carcinogen, damages liver, kidney, thyroid and immune system Adults—high blood pressure, digestive and nerve disorders, memory problems Acceptable levels in U.S. drinking water is 20 ppb	In United States, leaded gasoline, lead shot, leaded solder, and leaded paint banned in United States In older homes—replace copper plumbing that may contain lead solder and remove lead paint (house built before 1978) Remediate—Phytoremediation of soil with sunflowers
Mercury	Air, soil, water	Coal combustion, incineration, smelting Mercury in air settles into water where the inorganic mercury is methylated by bacteria to create the most toxic form methyl mercury Inorganic mercury salts may be in fungicides and disinfectants Human exposure from eating contaminated fish and shellfish—shark, swordfish, kingfish, and tilefish contain high levels of methyl mercury	Highly bioaccumulated and biomagnified Fish of particular concern are shark, swordfish, king mackerel and tilefish Inhibit frog metamorphosis Birds and mammals that eat fish also have high exposures Effects—reduced reproduction, slow growth and development, death	Acute—difficulty walking; loss of coordination; difficulty swallowing; tremors Bioaccumulates in many organs and directly damages cells Chronic—mutagen; teratogen, carcinogen; hallucinations; psychosis; irreversible brain damage Fetal exposure—mental retardation; attention disorders; seizures; blindness	Prevent—reduce release of mercury into atmosphere—EPA issued Clean Air Mercury Rule in 2005 to regulate emissions of mercury by utilities Federal bans on mercury in paint and pesticides Remediate—phytoremediation

Pollutant	Location	Source(s)	Environmental Impact	Human Health Effects	Prevent/Remediate
Methane	Air	Naturally produced during decomposition Anthropogenic sources are decomposition in landfills, wastewater treatment, and livestock production Emitted by the production and transport of fossil fuels	Greenhouse gas—20 times greater heat holding capacity than CO_2	Human health impacts related to global warming	Prevent—government regulatory action; use methane from decomposition as energy source
MTBE	Water	Gasoline additive that promotes complete combustion to reduce CO and O_3 release	Persistent in the environment, no studies indicate specific damage to wildlife	Not considered a threat to human health at this time; inhalation may be linked to lung cancer	Prevent—prevent leaking from underground storage tanks; use an alternative oxygenate in fuel Remediate—chemical processing of contaminated soil and water
NO_x	Air; water	Fossil fuel combustion—particularly gasoline; burning of solid wastes; industrial processes; Primary air pollutant that leads to photochemical smog N_2O from feedlots is a greenhouse gas Water pollutant derived from movement of fertilizers	Forms nitric acid (HNO_3) in atmosphere—contributes to acid deposition Directly damages cuticles on plants so damages leaves Reduces crop yields Reacts with sunlight to form photochemical smog Eutrophication when in the form of nitrites (NO_2^-) or nitrates (NO_3^-) Decreases visibility Low pH in aquatic systems may stress sensitive organisms; increases Al solubility and toxicity; decreases biodiversity by reducing food available at lower trophic levels	See impacts of ozone because NO_x forms ozone in photochemical smog Human health impacts from drinking excess nitrates Acute—children can have impaired oxygen transport; death may result Chronic—urinary problems, spleen damage, hypertension	Prevent—increase efficiency; decrease fossil fuel use; use fluidized bed combustion; emissions standards for motor vehicles and power plant government regulatory action Remediate—treat acidified ecosystems with lime

Pollutant	Location	Source(s)	Environmental Impact	Human Health Effects	Prevent/Remediate
Oil (petroleum)	Water	Natural release; leaks from oil rigs, tankers, and pipelines	Directly kills animals; suffocation of filter feeders; damage to habitat	Toxicity related to specific chemicals in petroleum	Prevent—government regulatory action; require double hulls on tankers Remediate—oil degrading microbes
Oxygen demanding wastes	Water	Sewage; wastes from food production; meatpacking plants	Decomposition of the wastes removes DO in water resulting in anoxia	Human health impacts related to pathogens carried in the wastes	Prevent—government regulatory action; increase oxygenation of contaminated water
Ozone	Air	O_3; Secondary air pollutant arising from NO_x and VOCs; major component of photochemical smog	Damages plant cells; interferes with food storage in plants; reduces crop yields and increases disease susceptibility Decreases visibility	Usually highest during the summer; respiratory problems; irritate respiratory system resulting in discomfort, coughing, and throat irritation; impairs respiration; exacerbates lung disease	Prevent—government regulatory action; improved fuel efficiency; alternative fuel methods for automobiles; mass transit, carpooling, bicycles
Particulates (PM)	Air	Solid and liquid droplets in air; soot, dust, soil, and smoke from erosion or combustion source	If densely accumulate on plant leaves, can impair photosynthesis If particles are acidic, contributes to acid deposition May increase nutrient levels in surface water Impairs visibility Can damage and stain stone, which decreases the aesthetic value of monuments and statues	Irritate respiratory tract, leading to coughing and difficulty breathing; exacerbates lung disease	Prevent—electrostatic precipitators, bag filter, cyclone collector; wet scrubber; government regulatory action; energy efficiency decreasing fossil fuel use
PCBs	Soil; water	Solids and oily liquids used as lubricants, fire retardants, hydraulic fluids, adhesives, transformer fluids; landfills; incineration	Highly persistent; bioaccumulate Wildlife—deformities; high mortality rates; impairs reproduction	Acute—acnelike skin eruptions; skin pigmentation; vision and hearing impairment; spasms Chronic—mutagen; carcinogen; teratogen; interferes with function of thyroid hormones bioaccumulates in liver, muscle, and fat	Prevent—government regulatory action; use alternative materials Remediation—High temperature incineration; bacterial bioremediation
Pesticides	Soil; water	Agriculture; urban areas; golf courses	DDT—Very persistent Birds—eggshell thinning; infertility Mammals—toxic to embryos and fetuses	Neurotoxin—increases neurotransmitter release increasing excitability of muscles; tremors; convulsions; death	Prevent—government regulatory action; use nonpersistent nonbioaccumulating pesticides; IPM

Pollutant	Location	Source(s)	Environmental Impact	Human Health Effects	Prevent/Remediate
Phosphates	Water	Fertilizer; sewage effluent	Eutrophication	None noted	Prevent—decrease nonpoint runoff from agricultural areas; ban detergents that use phosphates; government regulatory action; aerate contaminated water
Plastics	Water	Boat wastes	Harms wildlife by wrapping around their bodies	Production more dangerous than product	Prevent—government regulatory action; ban ocean dumping from ships
Radioactive isotopes	Air; soil; water	Naturally occurring; used in medicine, nuclear power, nuclear weapons, tracers	Increased genetic damage due to radiation	Acute—radiation sickness; death Chronic—cancers	Prevent—government regulatory action; use alternative materials
Radon	Indoor air; rock Produced by the natural radioactive decay of uranium	Naturally occurring radioactive gas in rock beneath homes; well water More frequent in buildings with foundations, basements and in airtight buildings Action level is 4 picocuries/L	None due to ventilation	Chronic—second leading cause of lung cancer in United States; carcinogen	Remediate—build houses without basements or slabs; improve ventilation; avoid building on hot spots; seal cracks
Sediment	Water	Deforestation leading to erosion Agricultural erosion Natural weathering of rock Mining	Increased water turbidity decreases photosynthesis due to decreased light penetration; sediment may cover benthic organisms; sediment may clog gills of filter feeders; may cover rocks in salmon spawning sites in streams	None—it is a water quality issue	Prevent—government regulatory action; sediment buffers when constructing; erosion control methods used in agriculture and mining

Pollutant	Location	Source(s)	Environmental Impact	Human Health Effects	Prevent/Remediate
SO_2	Air	Coal combustion; industrial processes; smelting; petroleum refineries; natural sources such as volcanoes	Acid deposition when combines with water vapor in the atmosphere to form sulfuric acid (H_2SO_4) Directly damages cuticles on plants so damages leaves; weathers carbonate rocks; acidifies ecosystems; reduces crop yield Low pH in aquatic systems may stress sensitive organisms; increases Al solubility and toxicity; decreases biodiversity by reducing food available at lower trophic levels Damages buildings, statues, and monuments May create sulfur haze which reflects sunlight resulting in global cooling Decreases visibility	Respiratory difficulty; exacerbates lung and heart disease	Prevent—wash coal; use anthracite (less S); increase energy efficiency; decrease fossil fuel use; fluidized bed combustion; coal liquefaction; coal gasification; use natural gas; lime scrubbers; pollution credits and other government regulatory action; decreases fossil fuel use and increase energy efficiency Remediate—treat acidified ecosystems with lime or ammonia

Pollutant	Location	Source(s)	Environmental Impact	Human Health Effects	Prevent/Remediate
Thermal pollution	Water	Power plant and industrial cooling	Temperature sensitive organisms may be killed. Hotter temperatures may interfere with reproduction, growth rates, and levels of biodiversity. Fish may require more food because they grow all year around in warmer waters. Dissolved oxygen levels decrease in warmer water	None	Prevent—use alternative energy resources that do not require cooling; cooling ponds or towers to cool hot water prior to being released into natural systems
VOCs	Air	Indoor—paints, building materials, cleaning supplies, permanent markers, copiers, ETS. Outdoor—component of photochemical smog; natural production from plants; vehicles		Acute—headache, nausea, eye, nose and throat irritation; CNS damage; difficulty breathing. Chronic—some cause cancer, such as benzene, due to DNA mutation	Prevent—Avoid products that release VOCs; throw away partly used containers; catalytic converter in cars; government regulatory action. Remediate—increase ventilation; remove source

Student Tips for Writing AP Environmental Science Free Response Essays

- Your essays will be graded by highly qualified teachers and professors trained to carefully assess your essay's quality. They follow a rubric, a grading system that has the scoring assignments for different points, which lists all possible viable answers to an essay to ensure each student is given the correct amount of points for their answers. Each essay is worth a maximum of ten points. Within those ten points there may be several ways to obtain the points. Do not attempt to guess which portion of the question will be worth the most. Answer each portion of the questions as they are given.

- Read all four questions before you begin. Make a brief outline of what you want to address before beginning writing. Use your question book for this process on exam day, not the book in which you are writing your answers. You do not need to answer the questions in the order they are presented, however, you must ensure that you begin the answers on the pages where the question occurs.

- Read the question carefully—note terms such as *describe, discuss, explain, compare,* and *design*. These words have different meanings, and you must ensure that you address the question properly. Students must learn to *describe* or *explain* properly, fully answering the question. If the question asks students to both identify and describe—they must do both, as many times the entire explanation will be worth one point for the identification and description.

- Students have a tendency to explain the cause of an environmental problem, but do not relate it to an effect and consequence. Frequently they are on the right track, but they do not continue their thought. For example, cigarettes themselves do not cause lung cancer, but *smoking* cigarettes is the likely cause of lung cancer. *Cause, effect, and consequence must always be linked.*

- Students tend to mix up the words ecological, environmental, and economic when answering questions.

- If students are asked to identify a pro and a con of an issue, they should never be the same idea and simply restated for the opposing argument. You must come up with genuinely separate pros and cons.

- Do not bother restating the question or writing an introductory paragraph. It wastes your time and does not impress the readers. It may also be misconstrued as your first of two answers. Just start your answer. Do not make a conclusion paragraph. You do not need to sum up your essay.

- Your opinion about an issue is irrelevant in most cases—address the question as asked. Students typically have problems generating good scientific arguments for or against a particular subject. An argument should be a series of sentences that supports an original premise, not whether you agree with the subject in question. Support your position with valid scientific evidence, and you will be answering the question properly.

- Pace yourself since you only have about 22 minutes per question—do not dawdle.

- Be sure to define every scientific term you use. Tossing out terms like "biomagnification" or "eutrophication" does not mean you know what you are talking about. Define it and throw in a quick example to elaborate to ensure that you receive credit.

- If you cannot recall a specific term, describe what you know and explain the concept being addressed in the question.
- Do not be repetitive, but develop your ideas completely. Most students fail to finish their thoughts, and thus do not get the total points possible in a section due to an incomplete answer.
- Be concise—do not regurgitate every thing you know about a topic.
- Be sure that you follow any instructions.
  - If you are asked to give *an* example, this means one.
  - If you are asked to give two examples, give two. *You will be graded on the first two, regardless of how many you give.* If the first two you list are incorrect, you will get no points.
  - Be sure to give complete answers, including examples of the two that you describe.
- Attempt to answer *all* questions.
  - You will not receive penalty for an answer unless it contradicts a previous one. You can only receive points for correct answers, so at least try!
  - If you have no idea exactly what the question is asking, try to address the *concept* being asked. For example, one exam asked for specifics about a particular law. If you don't know/recall specifics, explain what the law is and how it is applied. Frequently you will pick up points for your explanation.
- Often a question will incorporate ideas from several aspects of the class that at first glance may seem unrelated. For example, a question about pest management may address soil erosion. You must make sure that the whole question is carefully answered, showing that you understand the relationships in many different subject areas of the class.
- If you are asked to draw a graph, consider the following:
  - Graph the independent variable on X axis and the dependent variable on Y axis.
  - Put units on both axes and label the variables.
  - Give the graph an appropriate title.
  - Make sure to place proportional increments on both axes. Do not simply use the data given to plot the line. Create appropriate increments.
  - Plot points and draw line. See if question asks for points to be connected or a best fit curve.
  - If you need to make two lines, ensure that the reader is easily able to determine which line is which.
- You may be asked to design an experiment. Consider the following:
  - Make the experiment feasible, preferably a lab experiment as opposed to field work if possible.
  - Generate an hypothesis—null is fine.
  - Only test *one* variable. Be sure to give a specific variable to be tested. Terms like nutrient or pollutant are too vague. Pick a specific nutrient (nitrogen) or a specific pollutant (lead).
  - Be sure to define the test group and control group. Mention variables that will be standard or constant in both (soil, water, light, amounts, etc.).

> ➢ Explain how you will conduct the experiment and what materials you will need.
> ➢ Explain what you will be measuring (growth, death, survival, height, etc.). Do not measure the "health" of an organism, measure a specific parameter of its health.
> ➢ Explain what your dependent and independent variables will be.
> ➢ Explain how your data will be analyzed and what results you anticipate.
> ➢ State how your conclusion will be derived.
> ➢ Be creative, well organized, and knowledgeable about conducting laboratory exercises.

- Write an essay. Single words or short phrases do not answer the questions. No charts, diagrams, or outlines are acceptable alone. You may make a diagram, but it must enhance your written answer. Be sure to label it carefully.

- Frequently you will be asked to do calculations. *No calculators are permitted on the exam.* You will not receive credit for calculations unless all work is shown in the question and all units are shown. Put your work and answer in a factor/labeling format and you will be fine. Students have difficulty with calculations, particularly those using scientific notation and energy conversions. Review factor labeling/factor analysis skills to be adept at answering the essays. Use units throughout the calculations to get the correct units in the answer. Consider the feasibility of your numbers as answers on calculation questions. For example, a $50,000 gas bill for winter heating of a home is excessive, but 15 billion tons of coal in a power plant is not excessive

- Students often have difficulty distinguishing between a pollutant and an effect. For example, you may think that SO_2 causes global warming, but it does not. To prepare for questions regarding pollutants, understand and memorize a chart of the most common pollutants, citing their sources, environmental effects, adverse human health effects, and ways to decrease pollutants.

- Students should be able to not only describe environmental problems, but they should be able to hypothesize solutions that may include interpreting scientific data. When interpreting trends shown in a graph or chart, students should use all of the information provided and explain the observed changes carefully.

- Students may be asked to suggest a policy to rectify an environmental problem. This should be a policy that is feasible and is currently not in place. For example, taxing gasoline to discourage driving to decrease NO_x will not work, as gasoline already is taxed. But the tax on gasoline could be increased, which would discourage driving, which might decrease the pollutant.

- Be sure that you label each part of the question. Most have parts a, b, c, d and may have subparts i, ii, and iii. Be sure the reader knows which part of the question you are answering.

- Write only in blue or black ink.

- Write neatly. If the reader cannot read your handwriting, you cannot receive points.

- If you do not like what you have written, draw a line through that line and begin again. There is no need to excessively scratch out sentences.

- Do the best you can and do not give up. Keep writing until the proctor calls time!

Calculation Guidelines

- Be sure when you multiply numbers with zeros that you account for them in your answer. For example, students multiplying 200 by 600 may arrive at 1200 as the answer. This cannot be correct, because you are neglecting the zeros that should be taken into account. The correct answer is 120,000.
- Remember when dividing by decimals to move the decimal point to the right in the divisor to create a whole number and also to the right in the dividend the same number of places.
- Students would benefit greatly from using scientific notation. You are far more likely to be able to get an answer correct when multiplying 4×10^5 by 3×10^2 than if you try to multiply 400,000 by 300. Students will usually make errors with their number of zeros.
- Be certain that you use the following rules when calculating using exponents:
  - When adding two numbers the exponents must be the same value.

    - $4.2 + 3.6 \times 10^2 = 0.042 \times 10^2 + 3.6 \times 10^2 = 3.642 \times 10^2$

  - Only when bases are the same can you add exponents when multiplying and subtract exponents when dividing.

    - $(2 \times 10^8)(4 \times 10^3) = 8 \times 10^{11}$

Conversion Guidelines

- Most of the conversions in environmental science will deal with energy values, populations, or other simple algebraic functions. You usually do not need to know the conversion factors as they are typically given in the question. You must be able to do factor analysis to determine the correct answer.
  - How many seconds are in one day?

24 ~~hours~~ / day × 60 ~~minutes~~ / 1 ~~hour~~ × 60 seconds / 1 ~~minute~~ = 86,400 sec per day

Essay Component Practice

Design an experiment

A scientist observed that salamanders in a local pond in the Appalachians were declining in number. The scientist wished to determine if the snow melt from the top of nearby mountains, which occurred at the time the salamanders were laying their eggs in the stream, was contributing to the loss of the amphibians. The snow was acidic, the result of emissions from coal burning power plants. Design an experiment to test whether the amphibians were affected by the acidic snow melt.

Graphing practice
The United States has the highest growth rate of any developed country. The population continues to grow rapidly due to its high fertility rate and immigrants. The U.S. fertility rate is the highest it has been since 1971, at 2.13. Sixty-one percent of the population growth in the last decade is due to immigration. The population of the United States in 1900 was 76 million and in 1920 had risen to 106 million. By 1940 the population was 132 million and increased to 180 million by 1960. In the year 1980 the population was over 200 million for the first time, at 227 million. By 2000, the population increased to 282 million and is projected to reach 325 million by 2020.

Graph the data of the U.S. populations and draw a smooth curve connecting the points.

Math skills practice

1. There are 640 acres in one square mile. If a whitetail deer requires a range of 20 acres, how many deer may be found in one square mile?

2. Energy Star is an EPA program that promotes energy efficiency in appliances. An Energy Star dishwasher requires a temperature of 140°F to properly function. The water coming out of most natural gas hot water heaters is 120°F.

Given: 1 BTU is the amount of energy required to heat one pound of water 1°F

   1 gallon weighs 8 pounds

   1 therm of natural gas provides 100,000 BTU

   1 KWH produces 3,400 BTUs

   Dishwasher used 1 kwh per load for its mechanical function

a. Calculate the number of BTUs needed to heat the 6 gallons of water needed to run the energy-efficient dishwasher.

b. Calculate the number of BTUs required for the mechanical function of a dishwasher.

c. Calculate the total number of BTUs to run a load of dishes.

d. Calculate the therms of natural gas needed to use the dishwasher.

Answers

Design an experiment—4 points
Hypothesis: The acid snow melt caused the death of the larval amphibians.

Null hypothesis: The acid snow melt had no effect on the larval amphibians.
1 point for hypothesis

Experimental design—3 points
Must have a controlled experiment where only one variable is changed—**1 point**
For example: Expose varying numbers of eggs to different concentrations of acid
                Independent variable—amount of acid
                Dependent variable—% survival of eggs
                Description of constants water volume, number of eggs; temperature of
                water; DO etc.—**1 point**
                Correlation with other studies, discuss stats, data gathering, etc.—**1 point**

Graphing practice

Make a table to create an easy way of examining data. Then ensure that you have equivalent spacing on both X and Y axes. Be sure to label axes with appropriate labels. The x axis is labeled with the years and the y axis with the population amount. Plot the points and connect with a smooth curve.

1900 76 million
1920 106 million
1940 132 million
1960 180 million
1980 227 million
2000 282 million
2010 325 million

Math practice

1.   There are 640 acres in one square mile. If a whitetail deer requires a range of 20 acres, calculate the number of deer found in one square mile.

640 ~~acres~~ / mile x 1 deer / 20 ~~acres~~ = 32 deer in one square mile

2. Energy Star is an EPA program that promotes energy efficiency in appliances. An Energy Star dishwasher requires a temperature of 140°F to properly function. The water coming out of most natural gas hot water heaters is 120°F.

Given: 1 BTU is the amount of energy required to heat one pound of water 1°F
          1 gallon weighs 8 pounds
          1 therm of natural gas provides 100,000 BTU
          1 KWH produces 3,400 BTUs
          Dishwasher used 1 kWh per load for its mechanical function

a. Calculate the number of BTUs needed to heat the 6 gallons of water needed to run the energy-efficient dishwasher.

    8 lb / ~~gal~~ x 6 ~~gal~~ = 48 lbs

    Change in temperature – 140 – 20 = 20° change

    48 ~~lb~~ x 20~~°F~~ x 1 BTU / ~~°F lb~~ = 960 BTUs to heat the water

b. Calculate the number of BTUs required for the mechanical function of the dishwasher.

    1 ~~kWh~~ x 1 3,400 BTU / ~~kWh~~ = 3,400 BTU

c. Calculate the total number of BTUs to run a load of dishes.

    3,400 BTU to operate + 960 BTUs to heat water = 4,360 BTUs

d. Calculate the therms of natural gas needed to use the dishwasher.

    4,360 ~~BTUs~~ x 1 therm gas / 100,000 ~~BTU~~ = .04 therms natural gas

Practice Essays

1. Eighty percent of the earth's bauxite is recovered from surface mines, with the majority of the reserves found in the tropics. Sometimes there is no overlying rock layer and sometimes the rock layer covering the ore can be 70 or more meters thick. Bauxite contains 45 - 60% Al_2O_3, smelted to recover the desired aluminum, then used in cans, airplanes, and electrical wiring. Chemicals released during smelting include perfluorocarbons, carbon dioxide, sulfur dioxide, and polycyclic aromatic hydrocarbons. Approximately 1 ton of aluminum can be recovered from 4 tons of bauxite. It takes 1 ton of aluminum to make 60,000 cans. In 2004, approximately 120 billion cans were made with 51 percent of them being recycled.

a. Calculate how many tons of bauxite had to be mined to produce the 100.5 billion cans made in 2004, assuming all of the cans were made from virgin aluminum.

b. Calculate how much less bauxite would need to be mined if the average aluminum can contained 40 percent postconsumer recycled aluminum.

c. Describe two environmental consequences of mining bauxite and two environmental consequences of smelting the ore to recover the aluminum.

d. Propose one economic incentive to encourage people in a community to recycle their aluminum, and explain two drawbacks that a community would encounter when trying to institute a new recycling program.

2. Urban sprawl and conversion to monoculture pine plantations has decimated much of the native temperate deciduous forests in the southeastern United States. Tennessee has been increasing the size of its forests while timber harvest continues. Tennessee ranks first in the country in the production of hardwood flooring and pencils and second in the country for production of hardwood lumber. Wood is among the top five crops in the state. Seventy-eight percent of the state's trees are hardwoods, and most of the forests are mixed hardwoods, or oak/hickory forests.

a. Identify and explain two methods of harvesting hardwood trees.

b. Describe two environmental benefits of a prescribed burn.

c. Make an ecological or economic argument for the harvest of temperate deciduous forests, and make an ecological or economic argument against the harvest of these forests.

1. a. Calculate how many tons of bauxite had to be mined in order to produce the 100.5 billion cans made in 2004, assuming all of the cans were made from virgin aluminum. **2 points—one for setup and one for correct answer**

120×10^9 cans $\boxed{\dfrac{1 \text{ lb Al}}{6.0 \times 10^4 \text{ cans}}}$ $\boxed{\dfrac{4 \text{ tons bauxite}}{1 \text{ ton Al}}}$ = 8×10^6 tons bauxite

b. Calculate how much less bauxite would need to be mined if the average aluminum can contained forty percent postconsumer recycled aluminum. **2 points—one for setup and one for correct answer**

8×10^6 tons bauxite x .40 = 3.2×10^6 tons bauxite

c. c. Describe two environmental consequences of mining bauxite and two environmental consequences of smelting the ore to recover the aluminum.

1 point for correct description of an effect of mining linked correctly to an environmental consequence for the first two descriptions and 1 point for a correct description of an effect of smelting linked to an environmental consequence

Mining Effect	Consequence
Disturbed land	Habitat disruption, erosion, deforestation
Noise pollution	Interferes with behavior of wildlife
Increased tailings, spoil, gangue left in spoil banks	Erosion, weathering, no topsoil so vegetation cannot grow
Increased sediment pollution	Increase surface water turbidity; clog gills of aquatic animals;
Particulate pollution	Respiratory effects on animals

Smelting Effect	Consequence
Increased perflourocarbons, CO_2, PAHs	Increased greenhouse effect; global warming
Increased SO_2	Ecosystem acidification, fish kills, weakens trees
	SO_2 is also a global coolant; could result in lower temperatures
Increased fossil fuel use to heat ore	Increased carbon dioxide (NO_x or SO_2 depending on fuel type)
Dams for hydroelectricity	Habitat disruption, flooding, increased temp downstream, sediment buildup

d. Propose one economic incentive to encourage people in a community to recycle their aluminum and explain two drawbacks that a community would encounter when trying to institute a new recycling program. **3 points; one for incentive and two for drawbacks**
Incentive
Charge per pound of waste
Charge a deposit on aluminum items
Rebate for recycling

Drawback
Expense of instituting recycling program
Compliance
Education
Mining company opposition

2. a. Identify and explain two methods of harvesting hardwood trees. **4 points—one point for each ID and one point for each correct explanation**

Clearcutting—removes all trees in an area
Seed tree cutting—cut nearly all of the trees in an area; leave a few trees to reseed the area
Selective cutting—cut only the mature trees in an area
Creaming or high grading—take only the best trees in an area and leave the best
Whole tree harvest
Coppicing—cut off tree, leaving stump to regenerate tree

b. Describe two environmental benefits of a prescribed burn. **2 points**

Improves nutrient cycling
Decreases duff and debris to prevent forest fires
Improves wildlife habitat
Opens seeds and cones that only open when burned
Kills many disease organisms that infect trees
Removes exotic species
Clears forest floor to allow new plants to grow
Removes noxious weeds

c. Make an ecological or economic argument for the harvest of temperate deciduous forests, and make an ecological or economic argument against the harvest of these forests. 4 points—two points for each premise for or against harvest and one point each for additional supporting statements for each argument

Economic	
For	Against
Jobs	Ecotourism
Lumber for houses, furniture, etc.	Subsistence hunting
Paper production	
Environmental	
For	Against
Removes diseased trees	Maintains microclimate
Improves overall forest health if sustainable	Carbon sink
	Trees prevent erosion
	Maintains wildlife habitat

Practice Test I
Multiple-Choice Portion
90 minutes
This section has 100 multiple-choice questions. NO CALCULATORS PERMITTED!

In questions 1-24, answers may be used more than once, once, or not at all.

Questions 1-4 refer to the following species interactions.

  a. predation
  b. mutualism
  c. parasitism
  d. commensalism
  e. competition

1. a bear eats a blackberry for nutritional value. The bear eliminates the seeds of the blackberry in its feces and the blackberry seeds can grow in a new location.

2. the remora attaches to the underside of a shark and eats leftover bits of food that the shark does not consume

3. male bighorn sheep use their horns to illustrate dominance during mating rituals

4. the Nile crocodile will open its mouth to allow the small Egyptian plover bird to enter its mouth to clean its teeth of debris

Questions 5-7 refer to the components of the nitrogen cycle listed below.

  a. assimilation
  b. nitrogen fixation
  c. ammonification
  d. denitrification
  e. nitrification

5. process by which atmospheric nitrogen gets converted to ammonia

6. nitrogenous organic compounds are converted to simpler inorganic compounds such as ammonia

7. bacteria convert ammonium ions in a series of steps to nitrate and nitrate ions.

Use the following air pollutants to answer questions 8-11.

> a. sulfur dioxide
> b. carbon monoxide
> c. nitrogen oxide
> d. tropospheric ozone
> e. carbon dioxide

8. naturally released as living organisms carry out cellular respiration

9. pollutant released during incomplete combustion of fossil fuels or biomass; binds red blood cells and impairs oxygen delivery to tissues

10. pollutant that arises from gasoline combustion; may result in eutrophication

11. released naturally from volcanoes and can reflect sunlight back out into space

Use the following choices of waste treatment methods for questions 12-16.

> a. primary water treatment
> b. sanitary landfill
> c. tertiary water treatment
> d. waste incinceration
> e. secondary water treatment

12. marsh plants are used to remove excess nitrates and phosphates from wastewater

13. removes solids and large particles from wastewater

14. used for municipal waste; can release dioxins and heavy metals

15. biological process that removes the dissolved and suspended organic material from wastewater

16. most common method of disposing of municipal solid waste in the United States

Use the following sustainable agriculture methods to answer questions 17-20.

> a. sterile males
> b. mulching
> c. terracing
> d. agroforestry
> e. Bt

17. method that could be used to control insect pests and decrease erosion

18. method used to increase soil fertility while decreasing erosion

19. method used to plant crops on areas with mountainous terrain

20. method used to control butterfly and moth larvae

Use the following water quality tests to answer questions 21-24.

  a. turbidity
  b. salinity
  c. dissolved oxygen
  d. biological oxygen demand
  e. pH

21. dumping of organic waste into surface water would increase this measurement due to an increase in decomposition

22. measures the hydrogen ion concentration of water

23. this measurement would rise in surface water due to runoff from a construction site

24. this measurement must be high in streams for species such as trout and salmon to survive

Use the graph below to answer question 25.

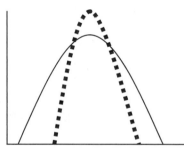

••••• New population
——— Old population

25. The above graph represents which type of natural selection?
a. stabilizing   b. directional   c. coevolution
d. disruptive   e. diversifying

26. Seven species of mustelid carnivore, including weasel, mink, otter, martens, and badgers, all coexist in the British Isles. Researchers have determined that they all feed upon different sized mammal prey as evidenced by skull length and length of canine teeth. This is an example of
a. competitive exclusion.   b. interference competition. c. resource partitioning.
d. mutualism.   e. territoriality.

Use the diagram below for question 27.

# of common
snook in
Charlotte Harbor

A B C D E

50° 60°   70°   80° 90°

Water temperature

27. Which of the following statements is correct?
a.   The range of tolerance is shown at D.
b.   Most of the organisms cannot survive in condition C.
c.   The optimal range for the organisms' survival is shown by letter B.
d.   The zone of physiological stress is shown at A.
e.   The zone of intolerance is shown at E.

28. Which of the following statements is correct regarding human impacts on the earth?
a.   Cultural eutrophication is frequently induced by runoff containing synthetic pesticides.
b.   Heavy metals released during smelting of ore rapidly break down in the environment.
c.   The building of highways and interstates has helped decrease habitat fragmentation.
d.   DDT and other organochlorine pesticides induce desertification.
e.   Deforestation of tropical dry forests and tropical rainforests is contributing to global
     warming.

Questions 29-31 refer to the diagram below.

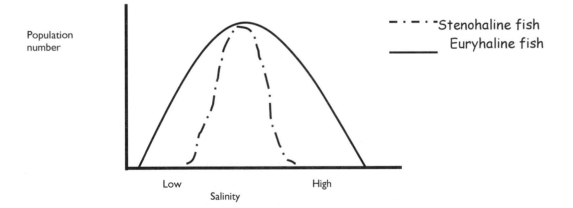

29. Stenohaline fish, such as goldfish, have tolerance for
a.   a broad range of depths.   b.   a broad range of temperatures.
c.   a narrow range of dissolved oxygen.   d.   a narrow range of salinities.
e.   lack of sufficient nutrients.

30. Flounder, a euryhaline fish, would likely be found in which of the following ecosystems
a. a stream.   b. an estuary.   c. a pond.
d. a large inland lake.   e. the open ocean.

31. Salmon are anadromous species that migrate from fresh water into salt water as
juveniles, then return to the fresh water to spawn. Salmon would be classified as
a. specialized stenohaline fish because they can move into fresh water.
b. euryhaline fish that become stenohaline fish prior to reproduction.
c. stenohaline fish that tolerate variable salinities.
d. euryhaline fish that spend part of their life cycle in fresh water and part in salt
water.
e. stenohaline fish because they must spawn in fresh water.

32. Which of the following biogeochemical cycles does not have a gaseous phase?
a. carbon   b. nitrogen   c. sulfur   d. water   e. phosphorus

Use the diagram for question 33.

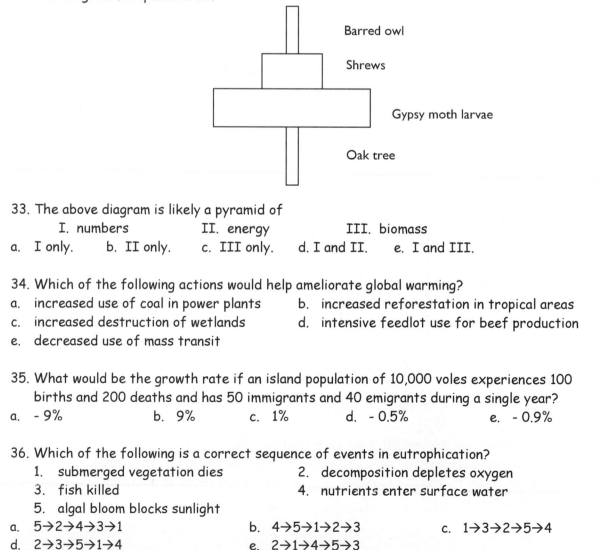

33. The above diagram is likely a pyramid of
          I. numbers   II. energy   III. biomass
a. I only.   b. II only.   c. III only.   d. I and II.   e. I and III.

34. Which of the following actions would help ameliorate global warming?
a. increased use of coal in power plants   b. increased reforestation in tropical areas
c. increased destruction of wetlands   d. intensive feedlot use for beef production
e. decreased use of mass transit

35. What would be the growth rate if an island population of 10,000 voles experiences 100
births and 200 deaths and has 50 immigrants and 40 emigrants during a single year?
a. - 9%   b. 9%   c. 1%   d. - 0.5%   e. - 0.9%

36. Which of the following is a correct sequence of events in eutrophication?
     1.   submerged vegetation dies   2.   decomposition depletes oxygen
     3.   fish killed   4.   nutrients enter surface water
     5.   algal bloom blocks sunlight
a. 5→2→4→3→1   b. 4→5→1→2→3   c. 1→3→2→5→4
d. 2→3→5→1→4   e. 2→1→4→5→3

37. Which of the following water diversion projects is currently undergoing restoration?
a. Lake Nasser, Egypt   b. Lake Mead, Nevada   c. Kissimmee River, Florida
d. Three Rivers Gorge, China e. Hoover Dam, Colorado

38. Clay has
a. good water holding capacity.   b. poor nutrient holding ability.
c. good permeability.   d. low porosity.
e. good aeration.

39. When particulates are inhaled in conjunction with sulfur dioxide, they aggravate the effect
    of sulfur dioxide on the lungs. This interaction is an example of
a. antagonism.   b. the additive effect.   c. synergy.
d. biomagnifications.   e. tragedy of the commons.

40. In a Florida sand pine ecosystem, 37 species of invertebrates disappeared in areas after
    the burrowing gopher tortoise was removed from the area. The gopher tortoise is most
    likely a(n)
a. community facilitator.   b. keystone species.   c. herbivore.
d. resource partitioner.   e. mutualistic organism.

Use the diagram of the results of an LC$_{50}$ study of copper sulfate to answer questions 41 - 43.

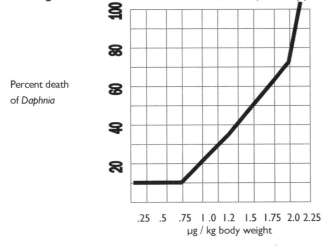

Percent death of Daphnia

.25 .5   .75 1.0 1.2 1.5 1.75 2.0 2.25
μg / kg body weight

41. What would be the threshold dose of the toxin?
a. 0.5 μg/kg   b. 0 .75 μg/kg   c. 1.0 μg/kg   d. 1.75 μg/kg   e. .0 μg/kg

42. hat would be the LC$_{50}$ for the toxin?
a. .75 μg/kg   b. .00 μg/kg   c. .2 μg/kg   d. 75 μg/kg   e. .25 μg/kg

43. What would be the best reason why these results could be applied to rainbow trout?
a. Rainbow trout might have a different physiological response to copper sulfate.
b. *Daphnia* and rainbow trout are aquatic, so they might have a similar response to a toxin.
c. Rainbow trout have a more complex anatomy than *Daphnia*, so they might respond similarly to a toxin.
d. Rainbow trout are more sensitive to lower oxygen levels than are *Daphnia*.
e. The liver of a rainbow trout might convert the toxin into a harmless substance.

44. What is the term used to describe a steel manufacturing plant using the waste heat from melting metals to create electricity?
a. eutrophication   b. remediation   c. cogeneration
d. subsidence   e. liquid metal fast breeder

45. Which of the following pieces of legislation might apply to draining a wetland in Florida that contains numerous waterfowl during the winter months?
    I. Endangered Species Act   II. Clean Water Act   III. Price-Anderson Act
a. I only   b. II only   c. I and II   d. I and III   e. I, II, and III

46. A population growing at a rate of 3 percent a year will double in how many years?
a. 3   b. 6   c. 12   d. 23   e. 33

47. If water quality tests on a stream detect high levels of fecal coliforms and nitrates, what is the most likely explanation?
a. The stream may be contaminated by industrial contaminants in the area.
b. The stream has been contaminated by runoff from a feedlot or pasture.
c. The stream is perfectly normal because all streams have high levels of coliforms and nitrates.
d. The stream has many fish and amphibians living in it, which increase the levels of coliforms and nitrates.
e. The stream has an effluent pipe from a properly functioning sewage treatment plant.

48. All of the following agricultural practices decrease soil erosion except
a. shelterbelts.   b. cover crops.   c. terracing.
d. conventional tillage.   e. applying mulch.

Questions 49-52 refer to the age structure diagrams for country X and Y in 2000 shown below.

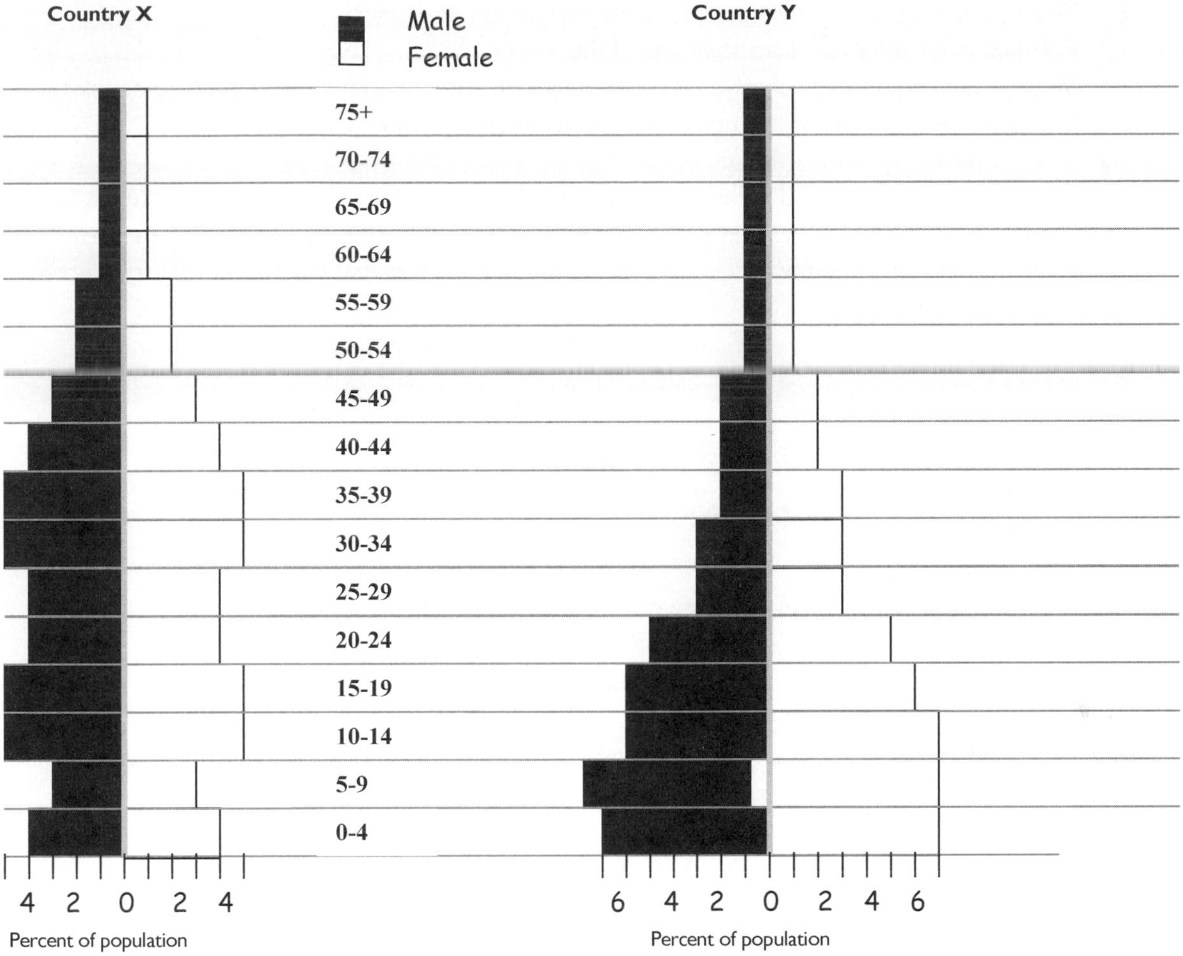

Country X

■ Male
□ Female

Country Y

75+
70-74
65-69
60-64
55-59
50-54
45-49
40-44
35-39
30-34
25-29
20-24
15-19
10-14
5-9
0-4

4 2 0 2 4
Percent of population

6 4 2 0 2 4 6
Percent of population

49. In country X approximately what percentage of the individuals were in the reproductive cohort under age 15?

a. 14   b. 18   c. 24   d. 34   e. 39

50. What inference cannot be made about country Y?
a.   The infant mortality rate is likely high in country Y.
b.   Country Y is in the preindustrial stage of demographic transition.
c.   Country Y is unlikely to be advanced in technology and medicine.
d.   Country Y is a developing country.
e.   Country Y's population is likely near 1 billion people.

51. In country Y, if the infant mortality rate declined while the birth rate remained steady, then the population would be expected to
a.   be more evenly dispersed among the age cohorts.
b.   be larger in the prereproductive cohort.
c.   remain the same as in the diagram in all of the cohorts.
d.   decline in the postreproductive cohort.
e.   increase in the reproductive cohort.

52. For country Y to stabilize their population, which of the following must occur?
a.   Infant mortality rates must decline by one-half.
b.   The death rate must decline in the post reproductive cohorts.
c.   Families must have no more than one child.
d.   Medical and technological advances must improve the life expectancy for the population.
e.   The country must lower the age at which women can marry.

53. When Walt DisneyWorld filled in acres of wetlands in south Florida to build theme parks
    and resorts, they had to build ponds and wetlands in other areas of their property to
    compensate for the damaged ecosystems. This type of environmental negotiation is an
    example of
a.   remediation.   b.   mitigation.   c.   restoration.
d.   reclamation.   e.   conservation.

54. All of the following gases are greenhouse gases except
a.   carbon dioxide.   b.   methane.   c.   water vapor.
d.   sulfur dioxide.   e.   nitrous oxide.

55. Which of the following is a source of radon found in homes?
a.   cigarette smoke   b.   a malfunctioning furnace   c.   mold growth
d.   the land under the house   e.   paneling and carpets

56. If a population is 1.2 billion and the growth rate is 2 percent, approximately how many
people would the population increase in a year if the growth rate did not change?
a. 2×10^3 b. 2×10^4 c. 2×10^5 d. 2×10^6 e. 2×10^7

57. Stratospheric ozone is important to life on earth because it
a.   absorbs heat from space.   b.   refracts visible light.
c.   binds reversibly to oxygen.   d.   reflects cosmic rays into space.
e.   absorbs UV light.

58. All of the following are environmental effects of thermal pollution except
a.   fish grow more slowly in warmer water.
b.   phytoplankton can carry out photosynthesis more quickly in the high temperatures and may
     cause an algal bloom.
c.   manatees seek out the warm water during winter months.
d.   shellfish grow more quickly and can be harvested in a shorter time for human consumption.
e.   small zooplankton may die of thermal shock due to the increased water temperature.

59. Which two factors are the most important for plant growth?
a.   high levels of dissolved oxygen and phosphorus
b.   adequate sunlight and water
c.   high levels of nitrogen and a low pH
d.   lots of wildlife and decomposers to cycle energy
e.   high levels of oxygen and sulfur

60. Which of the following best illustrates the environmental impact of an invasive exotic species?
   a. The loss of gopher tortoises in portions of Texas and Florida has led to a decline in the number of indigo snakes.
   b. Zebra mussels brought to the United States in ballast water have nearly caused the extinction of native clams in the Great Lakes.
   c. The removal of Asian kudzu from Georgia parks costs taxpayer's thousands of dollars a year.
   d. Since 1978 more than 1,200 power outages have occurred in Guam due to the brown tree snake, a native of New Guinea, crawling on power lines.
   e. The exotic pink hibiscus mealybug, which feeds on fruit trees, is preyed upon by an exotic species of parasitic wasp.

61. Acidification of surface water is harmful to the wildlife in the area because
   a. the increase in pH causes damage to the skin of amphibians.
   b. fish eggs cannot hatch in acidic water.
   c. the natural buffers in limestone protect the animals from the acid.
   d. the acid will increase the number of predators tolerant to the acid.
   e. the acidic water decreases the availability of aluminum, which animals need in high amounts to maintain homeostasis.

62. The half life of tritium (^{3}H) is 12 years. How many years must pass before a stored amount of tritium is considered safe?
   a. 12   b. 24   c. 48   d. 96

63. Which of the following removes carbon dioxide from the earth's atmosphere?
   a. aerobic respiration   b. fermentation   c. volcanic eruptions
   d. evapotranspiration   e. photosynthesis

64. All of the following methods could be used to cool a home naturally except
   a. using superinsulation and insulated windows.
   b. planting trees in front of windows or use awnings.
   c. using fans to circulate air.
   d. having a dark-colored roof rather than a light-colored roof.
   e. ventilating the attic to remove heat.

65. An area's topsoil is 1 m thick. Topsoil forms at a rate of 2.5 cm every 100 years, and the rate of erosion is 5 mm per year. Assuming the rates remain the same, how much soil will be present in 200 years?
   a. 100 mm   b. 5 cm   c. 100 cm   d. 1 m   e. 2 m

66. All of the following contribute to the endangerment of the Florida panther except
   a. captive breeding programs.   b. genetic bottlenecking.
   c. habitat fragmentation.   d. geographic isolation.
   e. low fecundity.

67. What would be the efficiency of energy transfer in an ecosystem in the absence of the second law of thermodynamics?

a. 0   b. 10   c. 20   d. 50   e. 100

Use the diagram to answer question 68.

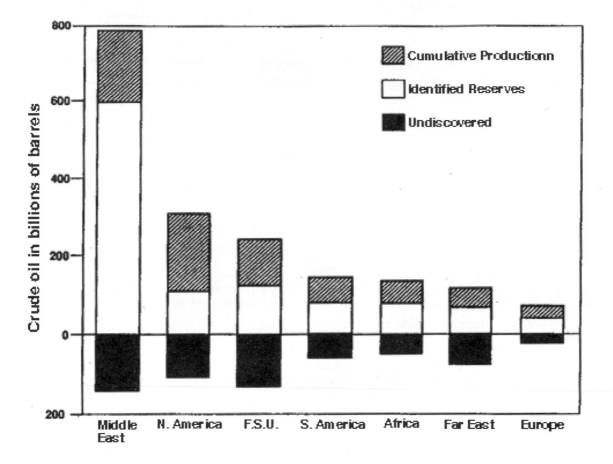

68. Which best describes the global distribution of crude oil as shown in the diagram?
a. The Middle East has the greatest amount of identified reserves.
b. The former Soviet Union has fewer reserves than Africa.
c. North America's reserves are greater than its production of crude oil.
d. The United States has most of the crude oil in North America.
e. South America produces the largest number of barrels in the world.

69. Which of the following is not an example of mechanical weathering?
a. a retreating glacier pulling till across a mountainside
b. frost wedging occurring in boulders
c. a swiftly flowing stream running over a riverbed of rock
d. ichens growing on the surface of a rock releasing carbonic acid
e. tree roots burrowing through cracks in rocks

70. Which of the following activities is not currently allowed in national parks?
a. camping   b. sport fishing   c. hiking   d. boating   e. sport hunting

71. Which of the following is an advantage of using oil shale and tar sands in lieu of conventional oil?
a. There are moderate existing reserves of the oil shale and tar sands.
b. There is no water pollution from the mining of the oil shale and tar sands.
c. They are relatively inexpensive to recover.
d. They result in little land degradation.
e. They do not emit carbon dioxide when combusted like conventional oil.

72. Which of the following international agreements focused on decreasing ozone depletion?
a. World Conservation Strategy
b. London Dumping Convention
c. Montreal Protocol
d. Kyoto Protocol
e. CITES

73. Which of the following has not been a human health impact as a result of the shrinking of the Aral Sea?
a. high maternal and infant mortality rates
b. high incidence of respiratory disease
c. increased schistosomiasis
d. increased level of diarrheal diseases
e. increased incidence of hepatitis and tuberculosis

The graph below depicts the mortality rates in the United States due to infectious diseases. Use it to answer questions 74-76.

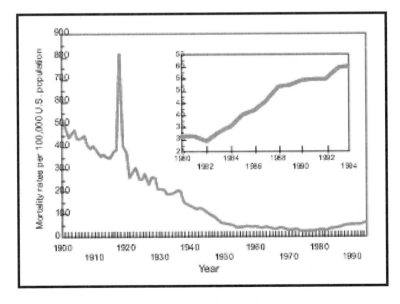

74. The spike in disease mortality from 1818-1819 was due to
a. yellow fever.
b. influenza
c. HIV/AIDS
d. malaria
e. cholera.

75. Deaths due to infectious disease increased by approximately what percent between 1982 and 1992?

a. 10   b. 30   c. 50   d. 75   e. 100

76. Although the United States has claimed to conquer infectious disease, the graph illustrates an increasing incidence of death due to infectious disease. All of the following have contributed to the increase in deaths due to infectious disease since 1982 except increased

a. incidence of HIV and AIDS.
b. spread of drug-resistant strains of tuberculosis.
c. death from septicemia from the use of IVs or abdominal surgeries and infections.
d. childhood vaccination in the United States.
e. number of deaths due to respiratory infections in the United States.

77. Subduction zones would exist at which of the following types of plate boundaries?

　　　I. divergent   II. convergent   III. transform

a. I only   b. II only   c. III only   d. I and II   e. II and III

78. An El Niño southern oscillation event would have which of the following environmental impacts?

a. surface water along the coasts of North and South America would become substantially cooler
b. increased rainfall on the western coast of the United States, causing mudslides and floods
c. increased rainfall in Australia leading to flooding during the normally dry seasons
d. increased hurricanes in the Southeastern United States
e. increased upwelling along the continents that increased the primary productivity

79. The greatest use of water in the United States is for

a. mining and municipal use.   b. livestock and public use.
c. aquaculture and industrial use.   d. cooling electrical plants and irrigation.
e. domestic and agriculture use.

80. One of the drawbacks to using photovoltaic cells to generate electricity is

a. fairly high net energy.
b. no carbon dioxide emissions because no combustion occurs.
c. systems can be large or small depending on the use.
d. low land use if used on roofs of preexisting buildings.
e. cells create DC current that must be converted to AC.

81. The best type of soil textures for agriculture would be

a. sand or clay.   b. clay or loam.   c. silt or loam.
d. silt or sand.   e. clay or silt.

82. Which of the following is an example of a density independent control on population growth?
a. a fungus called white pine blister rust kills all of the white pines in a southeastern forest
b. brucellosis disease causing death of elk, bison, and other ungulates in Yellowstone National Park
c. an increase in the population of wolves results in a decrease in the moose population in Canada
d. a honeybee queen bases her egg production upon food availability
e. a drought kills all of the sedges in a meadow

83. Which of the following would be the most sustainable fish to commercially harvest?
a. dolphin fish (mahi-mahi), which reaches sexual maturity in 4-5 months
b. blue fin tuna, which reaches sexual maturity at 4-5 years
c. spotted seatrout, which reaches sexual maturity in 1-2 years
d. pompano, which reaches sexual maturity at 2.3 years for males and 3.1 years for females
e. black fin shark, which reaches sexual maturity at 4-5 years for males and 6-7 years for females

84. Which of the following incorrectly describes mineral formation?
a. lava melting results in the formation of crystal, which forms minerals such as olivine
b. minerals, such as quartz, can form when crystallization occurs in hydrothermal vents
c. water may evaporate from a lake or seawater leaving behind mineral formations, like gypsum deposits
d. some minerals arise from metamorphism, including the iron ores of hematite and magnatite
e. calcium carbonate can be made when coral forms its outer shell

85. Which of the following would lead to sustainability of the global food supply?
a. increase irrigation in marginal lands to increase arable land
b. in arid regions, shift to plants that require less water such as wheat or sorghum
c. increase removal of surface water for irrigation
d. add more wells for irrigation in areas with ample groundwater
e. increase the amount of water used in irrigation to improve salinized soil on farmland

86. Which of the following countries has the greatest reserves of coal?
a. South Africa   b. Brazil   c. Canada   d. United States   e. Australia

87. Which of the following is an example of tragedy of the commons?
a. controlling stocking rates on federal range land
b. dumping low-level radioactive wastes into the ocean for disposal
c. setting size limits, bag limits, and seasons on grouper fishing in Florida
d. setting maximum allowable pollutant emissions from power plants
e. government establishing energy-efficiency standards for home appliances

88. Which of the following are expected to result from ozone depletion?
    I.   damage to crops and phytoplankton
    II. increased skin cancer and cataracts
    III. increase in sea level
a. I only   b. II only   c. III only   d. I and II   e. II and III

89. All of the following characteristics of an endangered species make it difficult to recover its population except when the animal
a. is harvested by humans for its hide or organs.
b. has a long gestation period.
c. requires a large range to feed itself.
d. requires a specialized diet.
e. requires a specific breeding or nesting area.

90. Which of the following is not a method to improve energy efficiency in a home?
a. add insulation in the attic and walls of a home
b. replace conventional windows with windows having low emissivity
c. use a tankless instant water heater to heat water for showers and baths
d. install a programmable thermostat
e. switch all lighting to incandescent light bulbs

91. Which of the following is considered a type of hazardous waste?
a. radioactive waste   b. oven cleaner   c. used motor oil
d. mining waste   e. fly ash from an incinerator

92. Which of the following agencies is correctly paired with its U.S. department?
a. Department of Agriculture—Bureau of Land Management
b. Department of the Interior—U.S. Forestry Service
c. Department of the Interior—Natural Resources Conservation Service
d. Department of the Interior—National Parks Service
e. Department of Agriculture—Environmental Protection Agency

93. All of the following are advantages of urbanization except
a. urban residents tend to have longer life spans and lower infant mortality rates.
b. recycling is more economically feasible because recyclables are concentrated.
c. urban residents have better access to medicine and education.
d. less land space is occupied by urban dwellers.
e. cities tend to have water supply and sanitation issues compared to rural areas.

94. Which of the following laws required the development of an environmental impact statement for any large construction project involving the federal government?
a. CITES   b. FIFRA   c. NEPA   d. CERCLA   e. SMCRA

95. The buildup of DDT in successive trophic levels is known as
a. bioremediation.   b. biomagnification.   c. accumulation.
d. eutrophication.   e. bioconcentration.

96. All of the following are acute effects of mercury exposure except
a. mental disturbances.   b. diarrhea.   c. respiratory difficulty.
d. vomiting.   e. kidney failure.

97. The leading contaminant of groundwater in the United States is
a. heavy metals.   b. coliform bacteria.   c. nitrates.
d. radioactive material.   e. petroleum products.

98. Which of the following might be a beneficial effect of global warming?
a. increased pests and crop disease in warmer areas
b. disappearance of some types of forests
c. flooding in low-lying areas and coastal zones
d. beach erosion
e. more precipitation in some arid regions

99. Which of the following is the preferred method to dispose of high-level radioactive wastes?
a. bury the wastes deep in the ground in geologically stable sites
b. put the wastes on a rocket and send it to outer space
c. place the wastes in a subduction zone
d. place the wastes in Antarctica
e. dump the wastes into the deepest part of the ocean in the Marianas Trench

100. Which of the following correctly describes how global circulation patterns form?
a. The sun heats the entire earth's surface evenly, thus causing the atmosphere to be evenly heated.
b. The tilt of the earth's axis at 23.5° and the rotation of the earth around the sun create the seasons.
c. The Coriolis effect causes the winds to be deflected left in the northern and southern hemispheres.
d. The sun is consistent in the amount of energy that it releases.
e. Cool air rises in the atmosphere and gains moisture, falling back to earth as warm wet air.

Practice Test I
Free Response
90 minutes

This section has four essay questions. NO CALCULATORS PERMITTED! Where calculations are required, be sure to show all work clearly. Be sure to thoroughly answer each question by using specific examples and support your answers with specific information. It is recommended that you use approximately 22 minutes to answer each of the following four essays.

1. The Hartwell Dam was build across the Savannah River in Georgia. The dam is about 7 miles from the junction of the Tugaloo and Seneca Rivers, which forms the Savannah River. The dam forms Hartwell Lake, which encompassed portions of both the Tugaloo and Seneca rivers. The Savannah River empties in to the Atlantic Ocean in Savannah, Georgia. The dam contains several electrical generators capable of producing 480 million kilowatt hours per year.

a. Calculate the number of homes which obtain their electricity from the Hartwell dam, if the typical home used 800 kilowatt hours per month.

b. Other than producing hydroelectric power, describe two economic or environmental benefits of building a dam across a river and two economic or environmental costs of building the dam.

c. Using an alternative energy resource such as hydroelectricity results in decreased use of conventional fossil fuels. Identify two air pollutants that would be emitted from a conventional fossil fuel burning power plant, and explain an impact that pollutant would have on human health or the environment.

2. Anatahan, one of the Marianas Islands just north of Guam in the Pacific Ocean, has a volcano violently erupting, covering the eastern portion of the island with lava and ash. Nitrogen is frequently a limiting factor to vegetation growth on islands. Nitrogen is essential for life because it is found in amino acids and nucleic acids. Molecular nitrogen (N_2) forms about 78 percent of the atmosphere by volume, however, this reservoir of nitrogen is not directly available to plants.

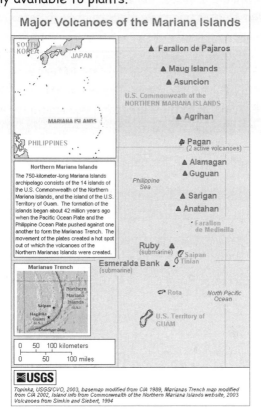

a. Volcanoes are dangerous not only for the lava they release, but the gases and ash that are released. Identify and explain two environmental consequences of an eruption.

b. Explain the type of ecological succession that would occur on Anatahan after a volcanic eruption.

c. Relative to the theory of island biogeography, discuss the biodiversity that would be found on the island.

d. Nitrogen is a limiting factor on islands. Explain two ways a farmer on nearby Saipan could sustainably increase the nitrogen in his or her soil to grow crops.

e. Discuss two ways humans could interfere with the nitrogen cycle, and give the environmental impact of the interference.

3. A recent study by the U.S. Geological Survey found that 97 percent of streams tested in agricultural areas and 61 percent of the shallow groundwater contained pesticides. One of the more common herbicides, atrazine, was found in many of the agricultural areas because it is used to kill broadleaf and grassy weeds on cropland. Atrazine is an endocrine disrupter in vertebrate animals. The maximum containment level goals set by the EPA for atrazine is 3 parts per billion. During rain events, levels in runoff have reached as high as 480 ppb. Frogs exposed to atrazine have altered growth rates, altered size, disrupted development, and increased susceptibility to disease. Amphibians are unique test animals because they must spawn in water and they respire through their skin to obtain oxygen. When tadpoles of varying species were exposed to differing concentrations of atrazine, a rather unusual pattern appeared. At 3 ppb, the mortality was 78 percent. At 25 ppb, the mortality was 45 percent and at 80 ppb atrazine, the mortality was 46 percent. Natural attrition of tadpoles was 18 percent death.

a. Graph the percent mortality of the tadpoles in response to the atrazine. Draw a smooth curve.

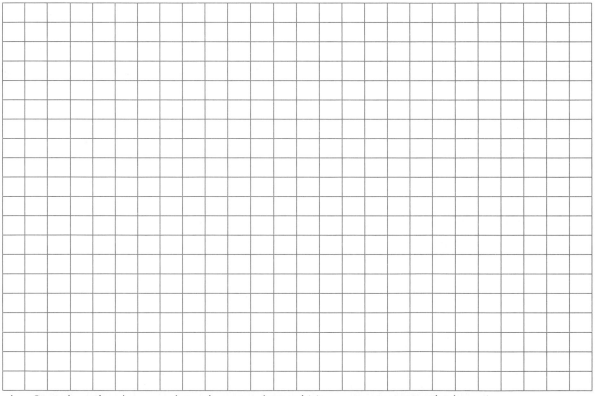

b. Based on the data, explain why a traditional LD_{50} experiment might have been inconclusive in these test organisms.

c. Explain what is meant by an endocrine disruptor. Identify a specific animal tissue or organ that might be affected by an endocrine disruptor.

d. Explain how these frogs have filled the role of an indicator species in these streams.

e. Identify another example of an indicator species, and explain what their presence or absence signifies in the environment.

4. Read the following article from the *Springfield Gazette*.

Springfield Gazette

═══════════════════════════════════════

Wolf Reintroduction in Yellowstone Yields Unexpected Results

The reintroduction of gray wolves to Yellowstone National Park has resulted in far greater environmental benefits than wildlife biologists could have initially predicted. Gray wolves are listed as endangered species in all of the lower 48 states except Minnesota, where they are considered threatened. The wolves were reintroduced between 1995 and 1996. The presence of the wolves appears to be increasing the numbers of young cottonwood, willow and aspen trees. Until recently, the large numbers of intensively grazing elk resulted in the small trees being devoured as saplings. There were virtually no new trees between the years of 1930 and 1990. Beavers have not been seen in the park since the 1950s, since without the trees, they had no food source. Their dams and ponds created wet areas for plants important for grizzly bear survival. With the wolves present in the park, the elk population has dropped from a high of around 20,000 to less than 10,000, and the trees have experienced a rebound. The first beaver dam in over 50 years has been seen along one of the rivers in Yellowstone. The increased vegetation has also decreased erosion along the riverbed. An additional effect has been seen on the numbers of coyotes in the park. Overall in the park the numbers of coyotes have dropped by 50 percent, and male coyotes are smaller now than when the wolves were missing from the park. The coyotes' prey, voles and mice which feed on seeds from grasses and trees, have experienced a population increase which has benefited the red foxes, hawks and eagles that live in the park. Grizzly bears also seem to be increasing, due to the large numbers of elk carcasses left by the wolves.

---

a. Diagram a food web with at least six members derived from the animals and plants listed in the article above.

b. Explain why the population of wolves in Yellowstone had declined so severely.

c. Explain three characteristics of endangered species that make them more prone to extinction.

d. Identify and describe two federal or international laws, treaties, or policies that prevent loss of biodiversity.

e. A population of 10,000 has 230 births per year, 60 deaths per year, 80 immigrants, and 40 emigrants. Calculate the growth rate in percent per year. Calculate the doubling time of the population.

Practice Test I Answers

<u>Multiple Choice</u>

1. b. Since both species benefit, this example illustrates mutualism.
2. d. Only the remora benefits and the shark is unaffected, illustrating commensalism.
3. e. The sheep are competing for mates.
4. b. Since both species benefit, the example illustrates mutualism.
5. b. The process by which atmospheric nitrogen gets converted to ammonia is called nitrogen fixation.
6. c. The process by which nitrogenous organic compounds are converted to simpler inorganic compounds such as ammonia is called ammonification.
7. e. The process by which ammonium ions are converted to nitrite ions and nitrate ions is called nitrification.
8. e. Carbon dioxide is naturally released as living organisms carry out cellular respiration.
9. b. Carbon monoxide is the pollutant released during incomplete combustion of fossil fuels or biomass. Carbon monoxide binds red blood cells and impairs oxygen delivery to tissues.
10. c. Nitrogen oxide arises from gasoline combustion and may result in eutrophication in aquatic ecosystems.
11. a. Sulfur dioxide is released naturally from volcanoes and can reflect sunlight back out into space, thus acting as a global coolant.
12. c. Tertiary water treatment uses marsh plants to remove excess nitrates and phosphates from wastewater.
13. a. Primary water treatment removes solids and large particles from wastewater.
14. d. Waste incineration is frequently used for municipal waste. Combustion of such waste can release dioxins and heavy metals.
15. e. Secondary water treatment uses bacteria to remove the dissolved and suspended organic material from wastewater, thus making it a biological process.
16. b. A sanitary landfill is the most common method of disposing of municipal solid waste in the United States.
17. d. Agroforestry, planting trees between or around different crops, is the only method listed that would both control insect pests and decrease erosion.
18. b. Mulching will naturally increase soil fertility while decreasing erosion.
19. c. Terracing is used to plant crops on areas with mountainous terrain
20. e. *Bacillus thuringiensis*, or Bt, is a biological control that will kill caterpillars, butterfly and moth larvae.
21. d. The biological oxygen demand would increase if organic waste was dumped into surface water due to an increase in decomposition. The DO would correspondingly decrease as a result of the increase in BOD.
22. e. pH measures the hydrogen ion concentration of water.
23. a. Turbidity is a measurement of the cloudiness of the water, or how much light could penetrate in water. This measurement would rise in surface water due to sediment runoff from a construction site.
24. c. The amount of dissolved oxygen must be high in streams for species such as trout and salmon to survive.

25. a. The graph illustrates stabilizing natural selection because the organisms become more average.
26. c. Since the carnivores all feed upon different sized mammal prey, it is an example of resource partitioning so that all of the animals can survive by having a slightly different food source.
27. e. The zone of intolerance is shown at E. The optimal range of survival is letter C. The zones of physiological stress would be B and D, and the zones of intolerance would be A and E.
28. e. Deforestation of tropical dry forests and tropical rainforests is contributing to global warming. Cultural eutrophication is frequently induced by runoff containing synthetic fertilizers. Heavy metals released during smelting of ore are persistent in the environment. The building of highways and interstates has helped increase habitat fragmentation. DDT and other organochlorine pesticides do not induce desertification.
29. d. Stenohaline fish can tolerate a narrow range of salinities, according to the diagram.
30. d. Euryhaline fish such as flounder, would survive well in an area of varying salinities such as an estuary.
31. d. All anadromous fish are technically euryhaline because they migrate from fresh water to salt water to live out their life and then return to fresh water to spawn.
32. e. The only biogeochemical cycle that does not have a gaseous phase is the phosphorus cycle, which makes it the longest of the cycles.
33. e. The diagram, because it is inverted, cannot be a pyramid of energy. It can be a pyramid of numbers and biomass, however.
34. b. Increased reforestation would create carbon sinks to remove carbon dioxide from the atmosphere, which would ameliorate global warming. Increased use of coal in power plants, increased destruction of wetlands, and decreased use of mass transit would exacerbate global warming due to increased carbon dioxide. Intensive feedlot use for beef production results in methane production, and methane is also a greenhouse gas.
35. e.
    $r = (b+i) - (d+e)$
        $= (100/10,000 + 50/10,000) - (200/10,000 + 40/10,000)$
        $= (10/1,000 + 5/1,000) - (20/1,000 + 4/1,000)$
        $= 15/1,000 - 24/1,000$
        $= -9/1,000 = -0.009 \times 100 = -0.9\%$
36. b. Given these choices, the correct sequence of events is as follows: 4. nutrients enter surface water → 5. algal bloom blocks sunlight →1. submerged vegetation dies → 2. decomposition depletes oxygen →3. fish killed
37. c. The Kissimmee River is being modified to its original flow in an attempt to restore the Everglades.
38. a. Clay has good water holding capacity, good nutrient holding ability, poor permeability, high porosity, and poor aeration.
39. c. When two substances have a greater effect than would be expected, it is called synergy.
40. b. The gophor tortoise is a keystone species because so many creatures depend upon its burrows for survival.
41. b. The threshold dose of the toxin is the first dose at which an effect is observed. The *Daphnia* begin to die off above natural attrition at 0.75 μg/kg.
42. d. The LC_{50} for the toxin would be 1.75 μg/kg because 50 percent of the *Daphnia* died at that dose.

43. b. The *Daphnia* and rainbow trout are both aquatic, and therefore they might have a similar response to a toxin. If the rainbow trout had a different physiological response to copper sulfate, then it would not be appropriate to extrapolate the data. The trout have a more complex anatomy than *Daphnia,* so they might respond differently to a toxin. Rainbow trout are more sensitive to lower oxygen levels than are *Daphnia* but that doesn't have anything to do with this experiment. If the rainbow trout's liver converts the toxin into a harmless substance, then the study would not easily be extrapolated.

44. c. Cogeneration is the term used to describe a steel manufacturing plant using the waste heat from melting metals to create electricity.

45. c. Both the Endangered Species Act and the Clean Water Act might apply to draining a wetland in Florida that contains numerous waterfowl during the winter months. The Price-Anderson Act protects the government and nuclear power plants from lawsuits in the event of an accident.

46. d. 70/3 = 23.33 – rule of 70s.

47. b. A stream contaminated by feedlot or pasture runoff might have high levels of fecal coliforms and nitrates. Industrial contaminants would not likely contain coliforms, and animals in a stream do not contribute to significantly high levels of nitrates and coliforms. If a sewage treatment plant is functioning properly the effluent would not contain coliforms.

48. d. Conventional tillage plows up the land, which displaces the soil, increasing erosion.

49. c. By adding the percentage of the population in the 0-4, 5-9, and 10-14 age range cohorts, you get a total of 24 percent.

50. e. You can infer that country Y likely has a high infant mortality rate because it has an age structure diagram shape indicative of a developing country. Country Y is likely in the preindustrial stage of demographic transition and is unlikely to be advanced in technology and medicine because it is a developing country. You cannot infer size of population based on these age structure diagrams.

51. b. In country Y, if the infant mortality rate declined while the birth rate remained steady, then the population would be expected to increase in the prereproductive cohort, because the overall number of children being born will now survive. The diagram will not be more evenly dispersed among the age cohorts, nor will all cohorts remain the same as in the diagram. The increased infant survival will not affect the postreproductive or reproductive cohorts for a generation.

52. c. To stabilize the growth, families must reproduce below replacement level fertility. If the infant mortality rates decline by one-half, the population will increase. The same is true if the death rate declines in the postreproductive cohorts and if the life expectancy improves. If the country lowers the marriage age, women are more likely to have even more children.

53. b. When ecosystems are rebuilt in a new location, it is called mitigation.

54. d. Sulfur dioxide reflects sunlight and thus cools the atmosphere.

55. d. Radon is a naturally occurring radioactive gas that arises from the land under a house.

56. e. You obtain the answer by multiplying the population times the growth rate.
    $$(1.2 \times 10^9) \times (2 \times 10^{-2}) = 2 \times 10^7$$

57. e. Stratospheric ozone is important to life on earth because it absorbs UV light.

58. a. Fish grow more quickly in warmer water since they are ectothermic.

59. b. Plants require sunlight and water to carry out photosynthesis.

60. b. The zebra mussels caused a decline of native species. Gopher tortoises are native to Florida and Texas. Although kudzu is an exotic, the problem described is an economic issue, as is the loss of power due to the brown tree snake. Using a parasitic wasp to destroy an exotic like the pink hibiscus mealybug is a method to remove an exotic.

61. b. Acidification of surface water is harmful to the wildlife in the area because fish eggs cannot hatch in acidic water. Acidified water causes a decrease in pH, not an increase. Natural buffers in limestone would prevent ecosystem acidification. The acid will not necessarily increase the number of predators. Acid increases the availability of aluminum, and animals do not need high levels to maintain homeostasis.

62. e. Radioactive elements are considered safe after 10 half lives. (10) (12) = 120 years

63. e. Photosynthesis removes carbon dioxide from the earth's atmosphere. Aerobic respiration, fermentation, and volcanoes increase carbon dioxide and evapotranspiration is the loss of water from a plant's leaves.

64. d. A dark roof would make a house hotter. In warm climates, a roof should be a light color.

65. b.

Erosion: 5 mm x 1 cm / 10 mm = .5 cm per year x 200 years = 100 cm erosion in 200 years
Increase topsoil: 2.5 cm / 100 years x 200 years = 5 cm in 200 years
Current topsoil: 1 m x 100 cm / 1 m = 100 cm
Current + topsoil formed - erosion of topsoil = 100 cm + 5 cm – 100 cm = 5 cm

66. a. Captive breeding programs are used to increase the population of Florida panthers.

67. e. The second law of thermodyamics states that all energy proceeds toward entropy, a measure of disorganization. Therefore in the absence of the second law, a system would be 100 percent efficient.

68. a. According to the diagram the Middle East has the greatest amount of identified reserves. The former Soviet Union has more reserves than Africa and North America's production is greater than its reserves. The diagram does not illustrate the crude found in the United States. The largest number of barrels in the world is produced in the Middle East.

69. d. Lichens secreting carbonic acid onto rock is an example of chemical weathering.

70. e. National parks allow all of the activities except sport hunting. Hunting does occur in some national preserves, lakeshores, seashores, rivers, and recreational areas, but does not occur in parks.

71. a. There are fairly good reserves of oil shale and tar sands. Water is polluted in their recover, they are quite expensive to recover, they require mining to remove, and they emit carbon dioxide when combusted.

72. c. The Montreal Protocol focused on decreasing ozone depletion by banning CFCs. The World Conservation Strategy refers to habitat conservation. The London Dumping Convention deals with oceanic dumping of radioactive, hazardous, and solid wstes. The Kyoto Protocol calls for a decrease in greenhouse gas production, and CITES regulates trade of endangered and threatened flora and fauna.

73. c. Increased schistosomiasis is associated with the building of the Aswan Dam in Egypt. All of the other health impacts have been seen in the area surrounding the Aral Sea.

74. b. The spike from 1818-1819 was due to the influenza pandemic.

75. c. The graph illustrates that the deaths were approximately 3 per 1,000 in 1982 and rose to 6 per 1,000 in 1992, an increase of 50 percent.

76. d. Increased childhood vaccination has decreased deaths from childhood diseases.

77. b. Subduction zones exist at convergent plate boundries when one plate is ocean and the other is continental.

78. b. An El Niño would result in increased rainfall on the western coast of the United States, causing mudslides and floods. The surface water along the coasts of North and South America would become substantially warmer. There would be drought in Australia. There are actually fewer hurricanes in the southeastern United States in an El Niño season. Upwelling is prevented by the warmer surface water.
79. d. The greatest use of water in the United States is for cooling electrical plants and irrigation.
80. e. One of the drawbacks to using photovoltaic cells to generate electricity is that the cells create DC current that must be converted to AC.
81. c. Silt and loam (a mixture of sand, silt, and clay) would be the best type of soil textures for agriculture.
82. e. Density independent variables affect the entire population equally—the size of the population is irrelevant. Therefore, the drought would kill the wildflowers regardless of the population size.
83. a. The dolphin fish mature quickly, so they are the easiest to harvest at a sustainable level with proper care.
84. a. When lava cools it results in the formation of crystal, which forms minerals such as olivine.
85. b. Shifting to plants requiring less water would increase the sustainability of the global food supply. All of the other choices require further use of irrigation, which is not sustainable.
86. d. Of these choices, the United States has the greatest reserves of coal.
87. b. Dumping low-level radioactive wastes into the ocean for disposal is an example of tragedy of the commons, whereas the other examples try to limit human impact on the environment.
88. d. Ozone depletion may cause damage to crops and phytoplankton and increased skin cancer and cataracts. Global warming is expected to increase sea levels.
89. a. That the animal is harvested for its hide or organs is not a characteristic of the species that makes it slow to recover.
90. e. Incandescent light bulbs are one of the biggest energy wasters in a home.
91. e. Fly ash from an incinerator must be disposed of in a hazardous waste landfill.
92. d. The BLM and the NPS are in the Department of the Interior. The USFS and the NRCS are in the Department of Agriculture. The EPA is a department itself.
93. e. Cities tend to have water supply and sanitation issues compared to rural areas and that is a drawback to urban areas.
94. c. The National Environmental Policy Act requires the development of an environmental impact statement for any large construction project involving the federal government. CITES is international law regarding endangered and threatened species. FIFRA regulates pesticides, CERCLA is the Superfund Act for hazardous waste site clean up and SMCRA regulates reclamation of mined lands.
95. b. The buildup of DDT in successive trophic levels is known as biomagnification.
96. a. The mental disturbances from mercury exposure are a chronic effect.
97. c. Nitrates are the leading contaminant of groundwater in the United States.
98. e. Increased precipitation in some arid regions might be a beneficial effect of global warming.
99. a. The preferred method to dispose of high-level radioactive wastes is to bury the wastes deep in the ground in geologically stable sites.
100. b. The tilt of the earth's axis at 23.5° and the rotation of the earth around the sun create the seasons. The sun does not heat the entire earth's surface evenly, and the Coriolis effect causes the winds to be deflected right in the northern and left in the southern hemispheres. The sun has variations in the amount of energy that it releases, and warm air rises.

Free Response

The answers given are some of the more common responses. Every possible answer to each question will not be listed.

1. a. Calculate the number of homes that obtain their electricity from the Hartwell Dam, if the typical home used 800 kilowatt hours per month. **2 points**

    480×10^6 kWh/year x 1 year/12 months x 1 month/800 kWh = 5×10^4 homes

    b. Other than producing hydroelectric power, describe two economic or environmental benefits of building a dam across a river and two economic or environmental costs of building the dam. **4 points**

Environmental	
Costs	Benefits
Loss of sediment downstream	Lake above dam provides new habitat
Loss of nutrients downstream	Lake provides space for migratory birds
Increased salinity of estuary	No thermal or air pollution
Floods ecosystem behind dam	Decreased erosion
Decreased water flow downstream	
DO below dam drops	
Increased evaporation from surface water	
Economic	
Costs	Benefits
Cost of dam maintenance/building	Controls flooding in floodplain
Displacement of human populations	Reservoir for humans for municipal and irrigation water
Increased disease	Ecotourism to lake
	Increasing hunting and fishing on lake

c. Using an alternative energy resource such as hydroelectricity results in decreased use of conventional fossil fuels. Identify two air pollutants that would be emitted from a conventional fossil fuel burning power plant, and explain an impact that pollutant would have on human health or the environment. **4 points**

Pollutant	Health / Environmental impact
Carbon dioxide	Global warming, ice caps melt, shifting biomes, and so on; acid deposition, ecosystem acidification
Carbon monoxide	Prevent blood from carrying oxygen; suffocation; headaches
Heavy metals (such as mercury from coal)	Appropriate to the heavy metal
NO_x	Acid deposition, ecosystem acidification; Precursor to photochemical smog—visibility; aggravates lung conditions
Particululates	Aggravates lung conditions; visibility decreased
Sulfur dioxide	Acid deposition, ecosystem acidification; sulfur haze; global cooling; aggravates lung conditions
VOCs	Aggravates lung conditions; decreased smog due to PANs or ozone

2. a. Volcanoes are dangerous not only for the lava they release, but the gases and ash that are released. Identify and explain two environmental consequences of an eruption. **2 points**

Eruption	Consequence
Lava	Damage to habitat, plants, and animals
Ash	Dust clouds block out light; dust enters surface water increasing turbidity and affects gills of filter feeders and fish
Sulfur dioxide	Acid deposition and acidification of ecosystems; acid aerosol that reflects sunlight to cool earth
Carbon dioxide	Greenhouse gas
Water vapor	Greenhouse gas
Hydrogen sulfide	Can suffocate animals

b. Explain the type of ecological succession that would occur on Anatahan after a volcanic eruption. **2 points**

The area would have been covered in lava and ash, so the land would undergo primary succession, with lichens the pioneer species. Lichens would help create soil, and then grasses, weeds, and shrubs would appear, followed by later successional species.

c. Relative to the theory of island biogeography, discuss the biodiversity that would be found on the island. **2 points**

Far away from mainland	Little biodiversity
Small island	Little biodiversity
Frequent damaging interference	Little biodiversity
Tropical climate	Would increase biodiversity slightly

d. Nitrogen is a limiting factor on islands. Explain two ways a farmer on nearby Saipan could sustainably increase the nitrogen in his or her soil to grow crops. **2 points**

Using inorganic fertilizer is not an option because the question specifically states sustainable agriculture. Any of the following methods would increase nitrogen in soil: mulch, green fertilizer, crop residue, feces from livestock, sewage sludge, compost, crop rotation, and so on with an explanation about how nitrogen would be increased in the soil.

e. Discuss two ways humans could interfere with the nitrogen cycle, and give the environmental impact of the interference. **4 points**

Fertilizer	Eutrophication; groundwater contamination
Auto exhaust; nitrogen dioxide	Eutrophication; acid deposition; photochemical smog
Livestock	Wastes contribute to water contamination
Forest and grassland conversion	Decreases N content of soil
Timber harvest	Decreases N content of soil
Tillage	Increases decomposition
Nitrogen oxide	Greenhouse gas

3. a. Graph the percent mortality of the tadpoles in response to the atrazine. Draw a smooth curve. Graph should have proper spacing on the axes and each axis should be properly labeled. **2 points**

All three data points should be plotted in the correct position. **1 point**
The natural attrition does not have to be plotted.

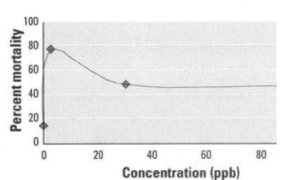

b. Based on the data, explain why a traditional LD_{50} experiment might have been inconclusive in these test organisms. **2 points**

The animals do not experience a threshold, so the results would not seem logical. Also, the traditional LD_{50} involves a dose response curve that has effects of the toxin magnify as the dose increases. Obviously in this case, the greater the dose, the less responsive the animal is to the toxin.

c. Explain what is meant by an endocrine disruptor. Identify a specific animal tissue or organ that might be affected by an endocrine disruptor. **2 points**

Endocrine disruptors are chemicals that interfere with the endocrine system or its hormones. Organs that might be disrupted in animals include thyroid gland, pancreas, pituitary gland, testes, ovaries, adrenal gland, pineal gland, parathyroid gland, and thymus.

d. Explain how these frogs have filled the role of an indicator species in these streams. **1 point**

They indicate that a change in the environment is occurring because they are not surviving. Their presence or absence indicated a specific environmental parameter.

e. Identify another example of an indicator species, and explain what their presence or absence signifies in the environment. **2 points**

Possible examples: benthic macroinvertebrates, song birds, top predators, lichens, other amphibians

A possible answer would be "Benthic macroinvertebrates, such as caddisfly and stonefly larvae, indicate good stream water quality."

4. a. Diagram a food web with at least six members derived from the animals and plants listed in the article above. **3 points one point for correct trophic level of animals; 1 point for direction of energy flow arrows; 1 point for having six members in food web.**

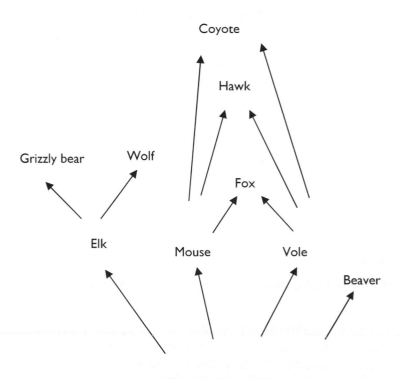

Coyote

Hawk

Grizzly bear   Wolf

Fox

Elk

Mouse   Vole

Beaver

Trees (willow, aspen, cottonwood)

b. Explain why the population of wolves in Yellowstone had declined so severely. **1 point**

Humans hunted them to extinction to prevent them feeding on wildlife and livestock and due to fear.

c. Explain three characteristics of endangered species that make them more prone to extinction. **3 points**

Any characteristic of a K-strategist, such as large body, long period of gestation, few offspring in lifetime, long time to reach sexual maturity, specialized feeding, specialized breeding area, and so on. One point for each characteristic as long as it includes why the characteristic makes them more prone to extinction.

d. Identify and describe two federal or international laws, treaties, or policies that prevent loss of biodiversity. **2 points**

Lacey Act	Prevents transport of living or dead animals or parts of animals across state lines
Migratory Bird Treaty Act/Migratory Bird Conservation Act	Protects species of migratory birds
Endangered Species Act	Prevents harassment, killing, harming of endangered or threatened species
Marine Mammal Protection Act	Prevents harm to marine mammals
Pittman-Robertson Act	Conserves wildlife and habitat
IWC whaling ban	Prevents whaling
CITES	Regulates international trade of endangered flora and fauna

e. A population of 10,000 has 230 births per year, 60 deaths per year, 80 immigrants, and 40 emigrants. Calculate the growth rate in percent per year. Calculate the doubling time of the population. **2 points—one for growth rate and one for doubling time**

$r = (b+i) - (d+e)$

$r = 220/10{,}000 + 80/10{,}000 \ - \ 60/10{,}000 + 40/10{,}000 =$

$r = 0.022 + 0.008 \ - \ 0.006 + 0.004$

$r = .030 - .010$

$r = 0.02 \ \times 100 = 2.0\%$ per year growth rate

Doubling time $= 70/r = 70/2\% = 35$ years

Practice Test II
Multiple Choice Portion
90 minutes
This section has 100 multiple choice questions. NO CALCULATORS PERMITTED!

In questions 1-24, answers may be used more than once, once, or not at all.

Use the following biomes to answer questions 1-4.

    a. chapparal
    b. boreal forest
    c. polar grassland
    d. desert
    e. tropical rain forest

1. temperatures below freezing most of year; permafrost
2. conifers are predominant vegetation; biome absent from southern hemisphere
3. periodic drought conditions; fire maintained ecosystem
4. extremely arid region; plant adaptations reduce water loss

Use the following legislation choices for questions 5-8.

    a. SMCRA
    b. National Wild and Scenic Rivers Act
    c. Price-Anderson Act
    d. Clean Water Act
    e. FIFRA

5. prevents building hydroelectric generating facilities in many areas with aesthetic value
6. limits liability of governments and companies in the event of a nuclear accident
7. forces companies to reclaim land disturbed by mining of coal
8. allows the EPA to determine safety of pesticide manufacture and use in the United States

Use the following energy choices for questions 9-12.

    a. biomass
    b. solar power
    c. nuclear fission
    d. coal
    e. natural gas

9. its use is responsible for 25 percent of all atmospheric mercury pollution in the United States
10. energy resource that is not a result of capturing the sun's energy
11. a power plant using this energy source would not require a water source for cooling

12. fewest pollutants released via combustion of all the fossil fuels

Use the following pollutants for questions 13-16.

>   a. formaldehyde
>   b. radon
>   c. tropospheric ozone
>   d. lead
>   e. mercury

13. second leading cause of lung cancer in the United States
14. common indoor air pollutant found in paneling, carpet, and furniture
15. air pollutant that contributes to photochemical smog
16. pollutant that can cause attention deficits and learning disabilities

Use the following pollution prevention mechanisms for questions 17-20.

>   a. electrostatic precipitators
>   b. lime scrubbers
>   c. fluidized bed combustion
>   d. bag filters
>   e. catalytic converters

17. designed to decrease sulfur emissions by burning coal with crushed limestone to form calcium sulfate
18. designed to remove NO_x from automobile emissions
19. decreases particulates in smokestack emissions by binding charged ash particles to a collecting plate
20. designed to remove sulfur dioxide from coal smokestack gases by precipitating the sulfur with a lime slurry to form a bottom ash

Use the following selections for questions 21-24.

>   a. desertification
>   b. salinization
>   c. deforestation
>   d. eutrophication
>   e. irrigation

21. tends to occur when overgrazing on marginal land
22. occurs when areas are clear cut at a rate beyond natural regeneration
23. occurs when lentic surface waters are contaminated with excess nutrients
24. occurs when water evaporates during irrigation of crops

25. A pyramid of biomass is most likely to have fewer producers than primary consumers in a(n)
a. boreal forest.   b. polar grassland.   c. estuary.
d. tropical rain forest.   e. tropical grassland.

26. The greatest volume of pollutant in surface water systems is usually
a. organic waste.   b. pesticides.   c. heavy metals.
c. sediment.   e. fecal bacteria.

27. The half-life of carbon-14 is 5,600 years. A fossil that has one-eighth the normal
    proportion of carbon-14 to carbon-12 is probably _____ years old.
a. 1,400   b. 2,800   c. 11,200   d. 16,800   e. 22,400

28. The presence of predatory starfish on a rocky shoreline holds in check the population of
    mussels, which would out compete many of the other sessile organisms that would attach
    to the rock surfaces. The role of the starfish in the community is that of a(n)
a. parasite.   b. keystone species.   c. indicator species.
d. exotic species.   e. flagship species.

Questions 29 and 30 refer to the diagram below.

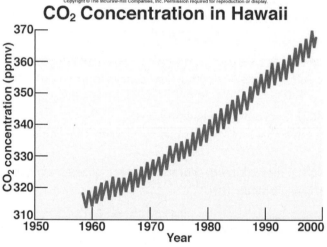

29. The data illustrated in the chart above is alarming because the increase in CO_2 is
    concurrent with an increase in global
a. sulfur dioxide emissions.   b. El Niño effects.   c. albedo effect.
d. temperatures.   e. alternative energy use.

30. Which of the following actions will potentiate the trend illustrated in the graph?
a. increased deforestation   b. increased mineral mining
c. increased use of wind energy   d. increased use of mass transit
e. decreased use of nuclear power

31. Nonutilitarian benefits of forests include all of the following except
a. aesthetic value of the redwoods found in the Pacific Northwest
b. banana slugs possessing intrinsic value as a part of nature
c. old growth forests in the Pacific Northwest serve as nesting sites for the northern spotted owl and the marbled murrelet
d. forests serve as carbon sinks
e. forests are the major source of fuelwood needed by people in developing countries

32. In Costa Rica cloud forests are drying up, likely due to global warming. As a result, lowland forest birds are being forced to extend their ranges up mountain slopes. The birds are serving as
a. exotic species.
b. consumers.
c. keystone species.
d. indicator species.
e. mutualists.

33. All of the following are reasons a prescribed burn might be used in a section of forest except
a. to eradicate the insect pest bark beetles in pine trees.
b. to remove the larger trees so that the smaller trees will have adequate sunlight.
c. burns allow regeneration of some species of pines whose cones only open when burned.
d. to remove exotic species such as the air potato from Florida scrub ecosystems.
e. buildup of woody debris in a southern pine forest.

Question 34 refers to the diagram below, which illustrates the sources of mercury emissions in the United States.

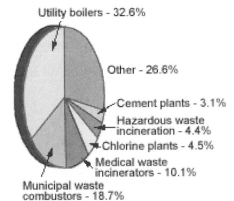

34. According to the chart, which of the following would reduce mercury emission in the United States the most?
a. place hazardous wastes into a specially designed sanitary landfill instead of incinerating the waste
b. use alternative energy resources such as wind or hydroelectric rather than fossil fuels to generate electricity
c. cease municipal waste combustion and put solid waste into open dumps
d. use more extensive pollution controls at cement plants
e. place more stringent regulations on the release of mercury from medical waste incineration

35. Which of the following contaminants would be unlikely in rural runoff?
a. heavy metals   b. oil   c. pesticides   d. fertilizer   e. sediment

36. Which of the following would be the method preferred by biologists for harvesting oak trees for the furniture industry?
a. creaming   b. shelterwood cutting   c. clear-cutting
d. seed tree cutting   e. coppicing

37. The type of coal preferred for use due to low sulfur and high heat release is
a. lignite.   b. peat.   c. anthracite.   d. bituminous coal.
e. sub-bituminous coal.

38. Disadvantages of wind energy include all of the following except
a. wind energy requires large amounts of land to locate the windmills.
b. wind farms must be located in areas with a steady and reliable wind availability.
c. windmills may kill migratory birds and raptors.
d. wind energy releases numerous pollutants during energy production.
e. windmills may be aesthetically undesirable in picturesque locations.

Questions 39 and 40 refer to the diagram below.

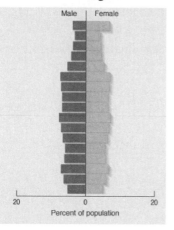

39. A country with an age structure diagram shaped as above likely has which of the following characteristics?
a. a population over 250 million people   b. a low GNP
c. a high infant mortality rate   d. be highly industrialized
e. little access to medicine and physicians

40. This country is likely to have
a. a low marriage age for women.
b. many children working menial jobs.
c. little availability of birth control.
d. lack of educational opportunities for women.
e. a successful pension system.

Use the paragraph to answer questions 41 and 42.
There are two species of barnacle that live on the coastline of Scotland. When the barnacle species *Balanus* is present in the intertidal zone, the *Chthamalus* barnacles live only in the upper intertidal zone. When the species *Chthamalus* is alone on the rocky coast they live in the lower portion of the intertidal zone.

41. The barnacles are exhibiting what type of competition?
a. predation   b. parasitism   c. intraspecific   d. interference   e. allelopathy

42. The niches in this ecosystem would best be described by which of the following statements?
a. *Balanus* can fill its realized niche in the presence of *Chthamalus*.
b. The presence of *Balanus* prevents *Chthamalus* from occupying its fundamental niche.
c. The presence of *Balanus* allows *Chthamalus* to its fundamental niche.
d. *Balanus* fills its fundamental niche when *Chthamalus* is present in the ecosystem.
e. *Balanus* cannot occupy its realized niche in the presence of *Chthamalus*..

43. If the efficiency of energy transfer in an ecosystem is 10%, and the grasses of the ecosystem obtain 1,000 units of energy from the sun, how many units will be available to feed the snakes that are feeing on the mice present in the ecosystem?
a. 100   b. 20   c. 10   d. 1   e. 0.1

Use the diagram to answer question 44.

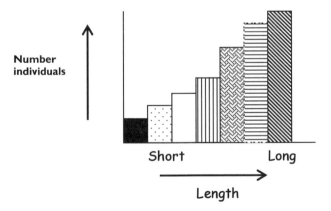

44. The graph illustrates scientific data collected on the length of bird beaks on a population of birds on a remote ilsand. The population is undergoing which type of natural selection?
a. artificial   b. stabilizing   c. disruptive
d. diversifying   e. directional

Use the diagram to answer questions 45-47.

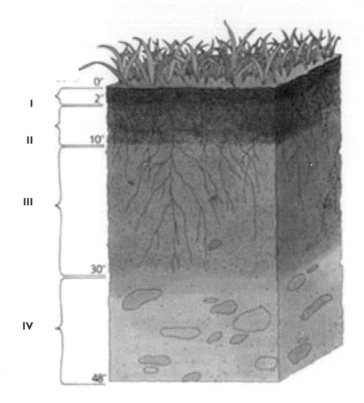

45. Which of the layer(s) of the soil profile above are most likely to be leached?
a. II only   b. III only   c. IV only   d. II and III   e. II and IV

46. Which of the layer(s) would contain the zone of illuviation or accumulation?
a. I only   b. II only   c. III only   d. IV only   e. I and II

47. Which of the layer(s) is also known as topsoil?
a. I   b. II   c. III   d. IV   e. Topsoil is not shown.

48. Which of the following will perpetuate global warming?
a. replanting deforested areas of Madagascar
b. increasing urbanization
c. increased use of mass transit
d. melting ice caps decreasing the albedo effect
e. release of sulfur dioxide from coal burning power plants

49. Which of the following characteristics of soil would most impede plant growth?
a. moderate permeability   b. low pH
c. high levels of nitrates and phosphates   d. loamy soil
e. numerous decomposers and high levels of organic material

50. The worst industrial accident of all time, which resulted in over 20,000 deaths, occurred in
   a. Three Mile Island, Pennsylvania
   b. Chernyobyl, Ukraine
   c. Love Canal, New York
   d. Prince William Sound, Alaska
   e. Bhobal, India

Questions 51-53 refer to the graph below.

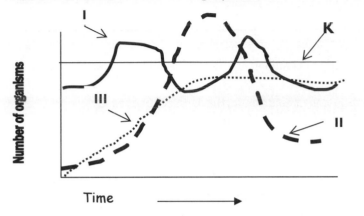

51. Population I is most likely a group of
   a. raccoons.   b. monkeys.   c. elephants.   d. mosquitoes.   e. deer.

52. The most likely explanation for the shape of the curve for Population II is that the population
   a. exceeded the carrying capacity of the ecosystem so quickly that environmental degradation resulted.
   b. reached ZPG when the death rate equaled the birth rate.
   c. reached its biotic potential due to favorable environmental factors.
   d. had no lag time in the population growth, which resulted in a slow population change.
   e. experienced linear growth, which caused the organisms to have high fertility rates over time.

53. Population III could be described as a(n)
   a. r-strategist who has irruptive growth.
   b. K-strategist with logistic growth.
   c. K-strategist with cyclic growth.
   d. r-strategist with linear growth.
   e. organism with a late loss survivorship curve experiencing irruptive growth.

54. Which method(s) are frequently used to disinfect sewage effluent in municipal systems?
      I. ozone   II. chlorine   III. UV light
   a. I only   b. II only   c. III only   d. I and II   e. I, II, and III

55. Which of the following best explains the location of deserts on earth?
   a. albedo effect
   b. rain shadow effect
   c. tilt of the earth's axis
   d. altitude of land where the desert is found
   e. rotation of the earth around the sun

56. Which of the following is most likely to occur when water is removed from a river for agricultural irrigation?
a. salt water intrusion
b. sinkholes in the surrounding land
c. subsidence in the associated watershed
d. increased salinity in the estuary downstream
e. increased phosphate levels in the river downstream

57. If a population of a country is 50 million and the growth rate is 2.0 percent what will the population number in 70 years?
a. 150 million   b. 200 million   c. 300 million   d. 400 million   e. 500 million

58. One of the primary sources of dioxin contamination in the United States is
a. emissions from bleaching pulp for paper.   b. nuclear power plant emissions.
c. use by farmers as a fumigant for fields.   d. combustion of coal.
e. emissions form smelting.

59. Integrated pest management includes all of the following methods except
a. using organochlorine pesticides to kill insect pests in crops.
b. releasing sterile males to breed with wild females.
c. using biopesticides such as Bt to kill caterpillars.
d. using organophosphate in low doses to reduce insect numbers.
e. releasing beneficial insects such as ladybugs or praying mantis to control pests.

The diagram below illustrates the sensitivity of aquatic species to acidified ecosystems. Use it to answer questions 60 and 61.

	pH 6.5	pH 6.0	pH 5.5	pH 5.0	pH 4.5	pH 4.0
TROUT	▨	▨	▨	▨		
BASS	▨	▨	▨	▨		
PERCH	▨	▨	▨	▨	▨	
FROGS	▨	▨	▨	▨	▨	▨
SALAMANDERS	▨	▨	▨	▨		
CLAMS	▨	▨				
CRAYFISH	▨	▨	▨			
SNAILS	▨	▨				
MAYFLY	▨	▨	▨			

60. Which of the species above is the most tolerant of an acidic environment?
a. trout   b. perch   c. frog   d. snails   e. crayfish

61. According to this diagram, which species would be the most likely to be used as indicator species for acidity?

a. perch   b. clams   c. salamanders   d. frogs   e. trout

62. Ecological benefits of wildlife include all of the following except

a. pollination.   b. matter cycling.   c. ecotourism.
d. pest control.   e. ecosystem homeostasis.

63. Which of the following is a greenhouse gas released by decomposing animal wastes?

a. carbon dioxide   b. oxygen   c. ozone
d. nitrogen dioxide   e. methane

64. Which of the following examples of harvesting fish is the most sustainable?

a. using trawl nets to harvest shrimp in an estuary
b. catching a bag limit of grouper using a hook and line
c. using a drift net to harvest salmon species
d. catching tuna using a purse seine net
e. catching sharks using long-lining

65. Which of the following is a primary cause of endangerment of giant pandas?

a. habitat disruption for human settlements
b. establishment of wildlife corridors
c. increased planting of bamboo
d. protecting pandas from international trade
e. captive breeding programs

The diagram below illustrates the salinity and temperature of the Chesapeake Bay estuary. Salt water becomes denser as the temperature decreases and increased salinity also causes the density to increase. Temperature has a greater effect on density than does salinity. Use the diagram to answer questions 66-68.

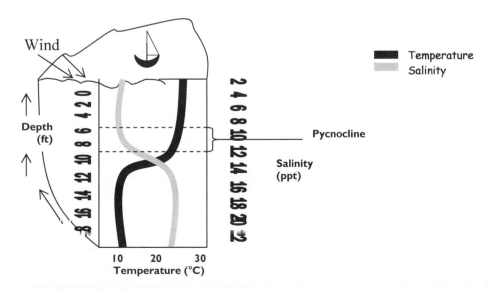

66. Based upon the information provided, you can infer that a pyncocline is
a. an area of the bay where waters of different densities are separated.
b. an area of the bay where the salinity is the greatest.
c. the section of the bay where photosynthesis is most likely to be carried out by phytoplankton.
d. an area where the water is the most dense.
e. the portion of the bay that has the highest temperature, thus making the water the least dense area of the bay.

67. Which of the following factors does not contribute to the mixing of the bay's waters?
a. wind blowing water away from the coastline
b. higher than normal tides
c. snow melt in spring
d. the Coriolis effect deflecting water out of the bay and toward the ocean
e. stratification of the bay throughout the summer

68. All of the following could be impacts on Chesapeake Bay as a result of global warming except
a. cooler bay temperatures.
b. increased precipitation.
c. loss of marshes and beaches.
d. increased flooding.
e. increased salinity.

69. Scientists tested the acute toxicity of dioxin on mice. The oral exposure LD_{50} was 114 μg/kg. Which of the following can be inferred from the data shown?
a. Rats, another type of rodent, should also have an LD_{50} of 114 μg/kg.
b. All mammals should have an LD_{50} of 114 μg/kg.
c. Fifty out of 100 mice exposed to dioxin at 114 μg/kg will die from the exposure.
d. Fifty percent of mice will become ill from exposure to dioxin, and then recover.
e. Humans will likely survive a dose of 114 μg/kg because they are so much larger than mice.

70. A major source of lead in the environment is
a. coal combustion.
b. smelting.
c. cigarette smoke.
d. volcanoes.
e. agricultural runoff.

71. All of the following are likely to produce volcanoes except
a. divergent plate boundaries.
b. convergent plate boundaries.
c. mantle plumes.
d. subduction zones.
e. transform plate boundaries.

72. To feed the rapidly increasing human population, which of the following programs would be the most sustainable?
a. develop more high-yield crops using genetic engineering
b. develop new pesticides that would increase crop yields
c. design new irrigation systems to use marginal land for agriculture
d. convert the land currently used for meat production into land for crop production
e. change the land ownership systems in developing countries to promote agriculture

73. One gallon of gasoline weighs approximately 6 pounds. The combustion of the gallon of gasoline releases approximately 20 pounds of CO_2. How many pounds of CO_2 are released if a car gets 25 miles per gallon and travels 500 miles?
a. 200 pounds   b. 300 pounds   c. 400 pounds   d. 500 pounds   e. 1.200 pounds

74. All of the following would help prevent soil erosion in semiarid regions except
a. using contour planting for planting drought-resistant crops.
b. managing stocking rates of cattle well below carrying capacity.
c. improving pastures by planting native species of plants.
d. using conventional tillage methods when planting crops each season.
e. using alley cropping to provide shade and help retain water.

75. Which of the following diseases is a waterborne illness?
a. cholera   b. tuberculosis   c. malaria   d. HIV/AIDS   e. avian influenza

76. Which of the following environmental impacts is not associated with broad spectrum pesticide use?
a. killing of nontarget species such as spiders and bees
b. bioaccumulation of pesticides in wildlife
c. cultural eutrophication of standing-water ecosystems
d. secondary pest outbreaks
e. evolution of pesticide-resistant species of insects

77. Which of the following is not a criteria pollutant according to the EPA?
a. sulfur dioxide   b. mercury   c. carbon monoxide
d. ozone   e. particulates

78. Which of the following is an example of closed-loop recycling?
a. adding crushed glass from bottles to asphalt for paving roads
b. recycling cardboard into writing paper
c. recycling plastics into plastic lumber for playground equipment
d. recycling steel from food cans to create a steel can
e. shredding tires to create a rubber mulch used as a groundcover on playgrounds

79. The prevailing surface winds found at the Arctic and Antarctic are called
a. westerlies.   b. doldrums.   c. tradewinds.
d. easterlies.   e. jet streams.

80. All of the following are considered benefits of large dams except
a. increased recreation in reservoirs.   b. consistent source of water.
c. producing hydroelectric power.   d. floor control.
e. decreased sediment flow to estuaries.

81. Which of the following is correct regarding a light water nuclear reactor?
a. The reactor uses a moderator such as graphite to slow the movement of neutrons.
b. The tertiary water circuit is used to cool the primary water circuit.
c. The secondary water circuit connects the primary and tertiary water circuits.
d. The primary water circuit is liquid and circulates to spin the turbine.
e. The reactor uses fuel rods packed with plutonium as its fuel source.

82. Which of the following minerals is correctly paired with its ore deposit?
a. hematite – copper   b. bauxite – aluminum   c. pyrite – gold
d. magnetite – zinc   e. quartzite – lead

83. Which of the following is an advantage of mass transit rail systems?
a. requires more land area than roads and parking lots for automobiles
b. decreases traffic congestion in cities
c. rail systems are expensive to construct and maintain
d. they are less energy efficient than cars
e. they are cost effective in nearly every part of the United States

84. Which of the following is not a feasible method to detoxify hazardous wastes?
a. Bacteria can be used to break down polychlorinated biphenyls contaminating soil.
b. Sunflowers can be used to removed lead from contaminated soil.
c. Toxic wastes may be incinerated to remove the majority of the toxins.
d. Bacteria can be used to remove uranium around weapons production sites.
e. Fungi can use extracellular digestion to break down MTBE in groundwater.

85. All of the following are correct examples of microclimate except
a. forests are cooler than a nearby meadow.
b. cities are warmer than surrounding rural areas.
c. tortoises in hot regions create burrows to help maintain their body temperatures.
d. sea breezes tend to heat coastal areas as compared to interior regions.
e. the Gulf stream brings warm air to make England warmer than its latitude would predict.

86. The General Agreement on Tariffs and Trade created the World Trade Organization, which governs world trade and makes binding decisions for member nations. Many environmental groups are opposed to the WTO because
a. globalization will raise health standards of developing countries.
b. reducing global trade barriers will benefit developing countries who suffer from trade agreements made to benefit developed countries.
c. economic growth in all countries will be stimulated by low prices of goods and services.
d. lower paying jobs are being eliminated in developed countries and moved to developing countries.
e. countries can set environmental standards on goods produced and sold in their country as well as on goods and services produced elsewhere.

87. Which of the following is a point source of pollution?
a. a smokestack from a coal burning power plant releases carbon dioxide and sulfur dioxide
b. waste from a commercial feedlot contaminates a stream
c. aerial spraying of pesticide on crops migrates to a nearby neighborhood
d. fertilizer runoff from a wheat field causes eutrophication in a catfish pond
e. after a summer shower, the storm water flows from the streets into a neighboring bay

88. When an aquifer is overdrawn in coastal areas, which of the following events may occur?
a. encroachment   b. salt water intrusion   c. subduction
d. aquifer transition   e. desertification

89. The dispersion pattern that you would expect of golden eagles would be
a. clumped.   b. random.   c. uniform.   d. aggregated.   e. range modified.

90. Which of the following is a result of filling in wetlands for construction?
a. filled areas serve as better nesting sites for migratory waterfowl
b. the filled area will be less susceptible to flooding
c. the loss of vegetation will result in decreased infiltration of water into aquifers
d. the land will be less polluted because the water runoff had so many pollutants
e. the area will be able to absorb more nitrates and phosphate from runoff once filled

91. The international agreement focused on climate change is
a. World Conservation Strategy.   b. London Dumping Convention.   c. Montreal Protocol.
d. Kyoto Treaty.   e. CITES.

92. All of the following are expected effects of ozone depletion except
a. decreased global warming.   b. increased photochemical smog.
c. loss of phytoplankton.   d. loss of amphibian populations.
e. food chain disruption in aquatic food webs.

93. Which of the following illustrate(s) natural selection?
    I. industrial melanism seen in peppered moths
    II. antibiotic resistance arising in strains of bacteria
    III. the stripes on a zebra make it difficult for a predator to pick out from the herd
a. I only   b. II only   c. III only   d. I and II   e. I, II, and III

94. Fossils are found in what type of rock?
a. igneous   b. sedimentary   c. metamorphic   d. basalt   e. slate

95. If a population of mice in a meadow exceeds their carrying capacity and experiences a
    population crash, what effect will the crash have on the carrying capacity?
a. no effect because carrying capacities are fixed
b. it will not affect the mouse carrying capacity but will lower the number of voles in the area
c. the environment will become degraded due to the population overshoot and the carrying
    capacity will be lowered
d. the carrying capacity will increase because the mice have died out
e. the carrying capacity will rise and then fall much like the population of mice

96. Which of the following is a nonrenewable resource?

a. trees   b. uranium   c. surface water   d. topsoil   e. crops

Use the graph below of China's total fertility rate to answer questions 97-99.

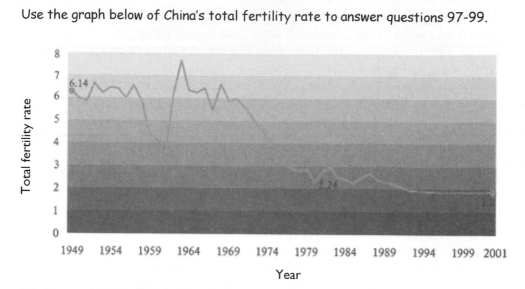

97. To reach ZPG, China's fertility rate must
a. decline to less than one.
b. remain slightly less than two.
c. take into account rural areas where children are needed as workers.
d. rise slightly above replacement level fertility.
e. return to the levels seen in 1949.

98. The diagram shows that China's total fertility rate dropped by what percentage between 1949 and 2001?
a. 10   b. 30   c. 50   c. 70   d. 100

99. All of the following effects were seen when China implemented its one child per family policy except
a. increased numbers of abortions.
b. increased use of birth control among Chinese women.
c. increased women in the workplace.
d. women waiting until they were older to have children.
e. lowered life expectancy in the elderly.

100. Which of the following is not a feature of a sanitary landfill?
a. geologically stable site   b. lined with clay and plastic
c. leachate pipes to remove contaminated water   d. methane recovery wells
e. refuse is placed in deep pits to allow natural decay

Practice Test II
Free Response
90 minutes

This section has four essay questions. NO CALCULATORS PERMITTED! Where calculations are required, be sure to show all work clearly. Be sure to thoroughly answer each question by using specific examples and support your answers with specific information. It is recommended that you use approximately 22 minutes to answer each of the following four essays.

1. The American Dust Bowl is portrayed as a natural disaster brought on by drought and high winds. However, poor agricultural practices played a significant role in the event. As a result of the Dust Bowl, the Natural Resources Conservation Service was formed to assist farmers in properly using their land, including teaching them about conservation tillage. Conservation tillage has many benefits, including reduction of soil erosion. Conservation tillage also has presented additional problems for farmers, including increased weeds and insect pests.

a. Using the soil texture triangle below, name the type of soil that would consist of 30 percent clay, 60 percent sand, and 10 percent silt.

b. Other than texture, describe another physical property of soil that would assist the NRCS in classifying the soil for its use.

c. Identify and discuss two methods of soil conservation that would be effective in areas with a high number of insect pests for crops.

d. Other than the soil erosion control methods that also work for pests, identify and describe two other examples of integrated pest management that could be used to help control crop pests.

e. Explain the difference between weathering and erosion, giving a specific example of each.

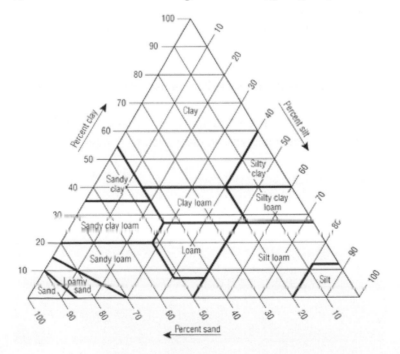

2. Read the following article from the *Springfield Gazette*.

Springfield Gazette

═══════════════════════════════════════

Should the United States follow Brazil's lead?

Brazil's military dictatorship was forced to ration gasoline due to the oil embargo of 1975. After the Iranian revolution in 1979, which drove oil and gasoline prices to all-time highs, the country did the unthinkable. They forced their automotive industry to switch from gasoline to ethanol engines. Brazil's ethanol is produced from a sugarcane by-product, bagasse. Brazil leads all countries with 4 million gallons produced in 2004. No regular gasoline is sold in Brazil at the pump; only gasohol of at least 24 percent ethanol or 100 percent ethanol is sold. Several auto makers, including Ford, GM, Daimler Chrysler, Mazda, Nissan, Mercedes, and Isuzu, currently sell cars capable of using E85, a mix of 85 percent ethanol with gasoline. The auto manufacturers in Brazil make flex fuel cars that can run on gasoline, mixtures of gasoline and alcohol, or 100 percent pure ethanol. The United States uses 140 billion gallons of gasoline a year, mostly from imported oil. There are drawbacks to using ethanol in American cars. Currently the United States has 95 ethanol plants that make only 4 billion gallons of ethanol in a year, primarily from corn. A typical car can get 25 miles per gallon on gasoline, but that car would need to burn about 1.5 times as much E85 as straight gasoline for the same mileage. Most likely, Americans will prefer gasoline for a long time.

---

a. Calculate the amount of gasoline a typical car would require to travel 500 miles.

b. Calculate how much E85 a flex fuel car would require to travel the same distance.

c. Assume that 10 percent of U.S. cars were flex fuel and used 100 percent ethanol, calculate the decrease in U.S. gasoline consumption. Determine if the current supply of U.S. ethanol is sufficient to meet the needed difference.

d. Using ethanol lowers the consumption of gasoline. Describe two environmental costs of switching to ethanol and two environmental benefits of switching to ethanol.

e. Propose one policy that the United States could implement to convince consumers to purchase flex fuel vehicles and give one drawback to the policy.

3. Modern agricultural techniques for raising livestock place animals in densely populated pens to rapidly increase their market weight with a specialized diet for increasing muscle and mass. Feeder pigs sell for $60 and weigh 80 lbs. Hogs are sent to market at 250 lbs. Hogs can be fed corn, barley, oats, or wheat to reach market weight. It takes approximately 3 lbs of feed to achieve 1 lb of body weight gain in a hog. In the United States, approximately 70 million acres of corn per year are grown for animal feed, with one acre of corn producing about 140 bushels of corn. One bushel of corn weighs 56 lbs, and 1lb of corn produces 280 kcal of energy. Humans require 2,000 kcal per day.

a. Calculate the amount of feed required to get a hog to market weight.

b. Calculate the number of humans that could be fed in a day on the corn to get one hog to market weight.

c. Calculate the total bushels of corn produced in a year in the United States.

d. Describe two environmental problems associated with feedlots and intensive animal production, and explain the impact on the environment.

e. Recently hogs have been selectively bred to have less fat and more muscle mass. Discuss one reason why the fat in pork was a human health concern.

4. To reduce solid waste in landfills, many local governments use incinerators to convert the waste to electricity. These plants generate 0.3 percent of the nation's electricity. These facilities also improve recycling rates, as the combustible material must be separated from the noncombustible components in the trash. These facilities must also monitor their ash, and any hazardous materials found in the ash will require that the ash be disposed of in a hazardous waste landfill.

a. Explain the difference between hazardous waste and municipal solid waste. Identify a specific example of each.

b. Describe three features of a municipal waste landfill.

c. Discuss two environmental problems associated with incineration of municipal solid waste.

Practice Test II Answers

<u>Multiple Choice</u>

1. c. The polar grassland, or tundra, has temperatures below freezing most of year and permafrost.
2. b. The boreal forest, or northern coniferous forest has conifers are predominant vegatation and is not present in the southern hemisphere.
3. a. The chaparral, or Mediterranean climate experiences periodic drought conditions and is therefore a fire maintained ecosystem.
4. d. The desert is an extremely arid region whose plants are adapted to reduce water loss.
5. b. The National Wild and Scenic Rivers Act prevents building hydroelectric generating facilities in many areas with aesthetic value.
6. c. The Price - Anderson Act limits liability of governments and companies in the event of a nuclear accident.
7. a. SMRCA - Surface Mining Reclamation and Control Act forces companies to reclaim land that has been disturbed by mining of coal.
8. e. FIFRA - Federal Insecticide, Fungicide, and Rodenticide Act allows the EPA to determine safety of pesticide manufacture and use in the U.S.
9. d. The use of coal is responsible for 25% of all atmospheric mercury pollution in the U.S.
10. c. Nuclear fission is the energy resource which is not a result of capturing the sun's energy. Biomass is the result of photsynthesis, and coal and natural gas are fossil fuels from prehistoric plant residues. Solar power is obviously derived from the sun's energy.
11. b. Solar power does not require water to cool the plant since no water seam turbines are used.
12. e. Combusion of natural gas releases the fewest pollutants of all the fossil fuels.
13. b. Radon is the second leading cause of lung cancer in the U.S.
14. a. Formaldehyde is a common indoor air pollutant found in paneling, carpet and furniture.
15. c. Tropospheric ozone is the air pollutant that contributes to photochemical smog.
16. d. Lead is the pollutant that can cause attention deficits and learning disabilities.
17. c. Fluidized bed combustion is the pollution control method designed to decrease sulfur emissions by burning coal with crushed limestone to form calcium sulfate.
18. e. A catalytic converter is the pollution control device designed to remove NO_x from automobile emissions.
19. a. Electrostatic precipitators decrease particulates in smokestack emissions by binding charged ash particles to a collecting plate.
20. b. Lime scrubbers are designed to remove sulfur dioxide from coal smokestack gases by precipitating the sulfur with a lime slurry to form a bottom ash.
21. a. Desertification tends to occur when overgrazing on marginal land.
22. c. Deforestation occurs when areas are clear cut at a rate beyond natural regeneration.
23. d. Eutrophication occurs when lentic surface waters are contaminated with excess nutrients.
24. b. Salinization occurs when water evacporates during irrigation of crops.
25. c. The estuary might have an inverted pyramid because the biomass of the rapidly reproducing phytoplankton is less than the zooplankton feeding upon it.
26. c. Sediment has the greatest volume of any pollutant in surface water systems.

27. d. One-half of the carbon-14 will decay in 5,600 years. One-quarter will decay in 11,200 years and one-eighth will decay in 16,800 years.

28. b. The starfish are keystone species because they contribute to controlling the population of another species.

29. d. The data illustrated in the chart is alarming because the increase in CO_2 is concurrent with an increase in global temperatures.

30. a. The more deforestation that occurs, the more carbon dioxide will accumulate in the atmosphere instead of the traditional carbon sinks of forests. The other factors will not contribute to an increase in carbon dioxide.

31. e. Nonutilitarian benefits are those that are not beneficial to humans as a resource. Using forests for fuelwood is a utilitarian benefit of forests. All other choices are non-utilitarian benefits of forests.

32. d. The birds are serving as indicator species because they are illustrating a change in the environment.

33. b. A prescribed burn should not burn the larger trees. It is designed to be a surface fire, not a fire that will take out mature trees.

34. b. By using alternative energy resources such as wind or hydroelectric rather than fossil fuels to generate electricity, the emissions, currently 32.6 percent, would decrease.

35. b. There are so few roads in rural areas, so one would not expect large amounts of oil in rural runoff. Heavy metals are frequently present in rural runoff naturally, and erosion increases sediment in runoff. Agriculture leads to erosion, and increases pesticides and fertilizers in runoff.

36. a. By taking only the best of the trees and leaving the rest of the trees to mature when creaming is employed, the ecosystem's integrity is maintained. Shelterwood cutting and seed tree cutting remove the majority of trees in an area, which would devastate animal habitat. Clear-cutting is not preferred because it increases erosion and results in the habitat being destroyed completely. Coppicing only works on trees that can come back from stumps after being cut. Oak trees are typically a species coppiced in the United States.

37. c. Anthracite is the type of coal preferred for use due to low sulfur and high heat release.

38. d. Wind energy releases no pollutants during energy production. All other examples are disadvantages of wind energy.

39. d. The shape of the age structure diagram is indicative of a developed country that would be highly industrialized. Population size cannot be determined from an age structure diagram. Developed countries have high GNPs, low infant mortality rates, and have access to medicine and physicians.

40. e. Developed countries typically have successful pension systems so that families do not need multiple children to care for their elderly parents. Developed countries have higher marriage ages, little to no child labor, available birth control and educational opportunities for women.

41. d. The competition is interference, because the *Balanus* interferes with the location of the *Chthamalus*.

42. b. *Balanus* inhibits *Chthamalus* from occupying its fundamental niche, because it is limited to its realized niche in the presence of the *Balanus*. *Chthamalus* does not fills its fundamental niche in the presence of *Balanus,* because it is confined to an area that

it does not prefer. It cannot be determined if *Balanus* fills its realized niche or fundamental niche in this ecosystem, since little information is provided about the *Balanus*.

43. c. If the efficiency is 10 percent, then 90 percent of the energy is lost at each trophic level.
    1,000 x 10% = 100 for the herbivores and 100 x 10% = 10.

44. e. The organisms are moving toward one extreme, so the population is experiencing directional selection.

45. a. Choice II is the A horizon, which is easily leached.

46. c. The zone of illuviation or accumulation of the leached material from the A horizon is the B horizon, designated letter III.

47. b. The A horizon is also known as topsoil.

48. d. If the ice caps are melted, then the albedo effect is decreased, which causes more sunlight to reach earth because it is not reflected back out into space. Replanting deforested areas of Madagascar will take up carbon dioxide, decreasing global warming. Increasing urbanization and increased mass transit would result in less carbon dioxide due to commuter traffic. Sulfur dioxide from coal burning power plants creates sulfur aerosols that cool the earth.

49. b. A low pH would most impede plant growth because plants do not do well in acidic soils.

50. e. The worst industrial accident of all time, which resulted in over 20,000 deaths, occurred in Bhopal, India. Three Mile Island and Chernyobyl were sites of nuclear accidents, Love Canal was a hazardous waste site, and Prince William Sound is where the Exxon Valdez had an oil spill.

51. d. Population I is most likely a group of mosquitoes, because the population rises and falls in a cyclic fashion.

52. a. The most likely explanation for the shape of the curve for Population II is that the population exceeded the carrying capacity of the ecosystem so quickly that environmental degradation resulted.

53. b. Population III could be described as a K-strategist with logistic growth. K-strategists remain near carrying capacity and the graph is of logistic growth, where the population experiences exponential growth until it reaches the carrying capacity, then levels off. Irruptive growth would have long periods of stability interrupted by periods of population explosions and crashes. Cyclic growth would be the graph shown in population I. Linear growth is not seen in populations.

54. e. All three methods are used to disinfect sewage effluent in municipal systems, because all three effectively kill bacteria and most parasites and inactivate most viruses.

55. b. The rain shadow effect best explains the location of many of earth's deserts. The albedo effect is the reflection of the sunlight into space off of earth's surfaces. The tilt of the axis and the rotation of the earth around the sun help explain the seasons. The altitude has very little to do with desert formation.

56. d. When water is removed from a river for agricultural irrigation, the most likely of the side effects would be increased salinity in the estuary. Salt water intrusion, sinkholes, and subsidence are usually the result of overdraw of groundwater. Phosphate levels should not be affected.

57. b. If a population of a country is 50 million and the growth rate is 2.0 percent the population will be 200 million in 70 years.
Applying the rule of 70, 70/2.0 = 35 years. Therefore 70 years would be two doubling times. 50 + 50 = 200 million for the second doubling.

58. a. One of the primary sources of dioxin contamination in the United States is bleaching wood pulp to make paper.

59. a. Integrated pest management uses a small amount of nonpersistent pesticides. They would not use organochlorine pesticides because they are highly persistent.

60. c. According to the diagram, the frog is the most tolerant of an acidic environment because it can withstand a pH of 4.

61. b. According to this diagram, the clams would be the most likely to be used as indicator species for acidity because they cannot survive a pH less than 6.

62. a. Ecotourism is a utilitarian benefit of wildlife, not an ecological benefit.

63. e. Methane is a greenhouse gas released by decomposing animal wastes.

64. b. Using a hook and line to catch fish is the most sustainable method. Using trawl nets damages the ocean floor and results in bycatch. The drift nets, purse seine nets, and long-lines also have a lot of bycatch and tend to catch fish below size limits and above bag limits.

65. a. Habitat disruption for human settlements is a primary cause of endangerment of giant pandas. All of the other factors would help them re-establish their population.

66. a. Based upon the information provided, you can infer that a pyncocline is an area of the bay where waters of different densities are separated. The salinity is greatest at the bottom of the water column, not the top. The photosynthesis must take place in the euphotic zone. The water is most dense at the bottom of the estuary. The pycnocline is not the area of the bay with the highest temperature.

67. e. Stratification prevents mixing of the bay's waters.

68. a. The water in the bay will likely increase in temperature, not decrease.

69. c. Fifty out of 100 mice exposed to dioxin at 114 μg/kg will die from the exposure. Rats would not necessarily have the same LD_{50} as a mouse, and the same is true of all other mammals. LD_{50} studies are lethal dose studies, in which the animals die, not just become ill. The dose is based upon body weight, therefore for humans to receive a dose of 114 μg/kg they will be exposed to more toxin than a mouse. It is not safe to say that they will be more resistant to the dose than a mouse.

70. b. Smelting is a major source of lead in the environment.

71. e. Transform plate boundaries are not likely to produce volcanoes—usually they result in earthquakes.

72. d. To feed the rapidly increasing human population, the most sustainable program would be to convert the land currently used for meat production into land for crop production. All of the other methods are energy-, land-, and time-consuming methods.

73. c. 400 pounds.
(20 lb CO_2/1 gallon gasoline) $\times$ (1 gallon gasoline/25 miles) $\times$ 500 miles = 400 lb CO_2

74. d. Conventional tillage plows up the soil and increases erosion.

75. a. Cholera is transmitted via water. TB requires person to person contact. Malaria is a mosquitoborne illnesss. HIV/AIDS is transmitted by blood or sexual contact. Avian influenza so far has passed only from infected birds to humans.

76. c. Cultural eutrophication of standing-water ecosystems is associated with use of inorganic fertilizers or runoff from animal or human wastes.

77. b. Mercury is not a criteria pollutant according to the EPA. The six criteria pollutants are sulfur dioxide, carbon monoxide, ozone, particulates, nitrogen dioxide, and lead.

78. d. Closed-loop recycling is when a substance is recycled and the same substance is remade. Therefore, recycling steel from food cans to create a steel can would be closed loop.

79. d. The prevailing surface winds found at the Arctic and Antarctic are called easterlies.

80. e. Decreased sediment flow to estuaries is a drawback of large dams.

81. a. The primary water circuit is steam and it heats the secondary water circuit, which spins the turbines. The secondary water circuit is cooled by the tertiary water circuit. The three circuits remain discreet. A reactor's fuel rods contain uranium pellets. The uranium is enriched to have about 3 percent U-235. The U-235 can only catch slower moving neutrons, and if the neutrons are moving quickly the U-238 in the pellets will catch them. Therefore, the reactor uses a moderator such as graphite to slow the movement of neutrons so that the U-235 can undergo fission.

82. b. Bauxite is aluminum ore. Hematite and magnetite are iron ore. Pyrite is sometimes called fool's gold, but contains iron. Quartzite is not a metallic mineral.

83. b. Mass transit rail systems decrease traffic congestion in cities. They require less land area than roads and parking lots for automobiles. They are expensive to construct and maintain, but are more energy efficient than cars. They are not cost effective in every part of the United States because you must have a large population for them to be cost effective.

84. e. At this time, there are no fungi that can break down MTBE. All of the other methods are being used.

85. d. Sea breezes cool coastal areas—they do not heat them up.

86. d. The General Agreement on Tariffs and Trade created the World Trade Organization, which governs world trade and makes binding decisions for member nations. Many environmental groups are opposed to the WTO because lower paying jobs are being eliminated in developed countries and moved to developing countries.

87. a. A smokestack from a coal burning power plant releases carbon dioxide and sulfur dioxide is a point source of pollution.

88. b. When an aquifer is overdrawn in coastal areas, salt water intrusion, where salt water is pulled into the freshwater aquifer, may occur.

89. c. The dispersion pattern that you would expect of golden eagles would be uniform.

90. c. The loss of vegetation will result in decreased infiltration of water into aquifers when wetlands are filled for construction. The filled areas destroy nesting sites for migratory waterfowl. The filled area is more susceptible to flooding and the land will be more polluted because the water runoff cannot pass through a wetland. The area can absorb more nitrates and phosphates as a wetland than when it is filled.

91. d. The international agreement focused on climate change is the Kyoto Protocol. The World Conservation Strategy refers to habitat conservation. The London Dumping Convention deals with oceanic dumping of radioactive, hazardous, and solid wastes. The Montreal Protocol focused on decreasing ozone depletion by banning CFCs. The Kyoto Protocol calls for a decrease in greenhouse gas production, and CITES regulates trade of endangered and threatened flora and fauna.

92. a. Global warming will likely increase as an effect of ozone depletion of more UV radiation reaching the earth's surface.
93. e. All are examples of natural selection.
94. b. Fossils are found sedimentary rock.
95. c. If a population of mice in a meadow exceeds their carrying capacity and experiences a population crash, the environment will become degraded due to the population overshoot and the carrying capacity will be lowered. Carrying capacities are not fixed and the mouse population would be affected. If the environmental degradation occurred it would likely deplete the voles but in addition to the mouse population. The environment will have to recover before the carrying capacity can increase. The carrying capacity will not rise due to the damage.
96. b. Uranium is a nonrenewable resource because it exists in a finite amount in the earth's crust and its supply is exhaustible.
97. b. The fertility rate must be less than replacement level fertility to reach ZPG. If the fertility rate is less than one, then the population will have too few young people to support the graying population. The fertility rate does not take into account jobs; it is the total number of children born to a female in her lifetime. The fertility rate cannot be above two or the population will continue to grow and if it returns to the levels seen in 1949, there will be a huge population boom in China.
98. c. The fertility rate in 1949 was 6.14 and the rate in 2001 was 1.8, a decline of 70 percent.
6.14 - 1.8/6.14 x 100 = 70.68
99. e. The life expectancy of the elderly did not change.
100. e. The refuse is covered with layers of soil to keep vermin and fires/explosions from the waste site.

Free Response

The answers given are some of the more common responses. Every possible answer to each question will not be listed.

1. a. Using the soil texture triangle below, name the type of soil that would consist of 30 percent clay, 60 percent sand, and 10 percent silt. **1 point**
    The soil is sandy clay loam.

   b. Other than texture, describe another physical property of soil that would assist the NRCS in classifying the soil for its use. **2 points—one for describing soil and an additional point if describe the way the information is used**

    Permeability of the soil, or how quickly water flows through the soil
    Porosity, or the water holding ability of the soil
    Color of soil—dark soil is indicative of high nutrients; iron soils are yellowish-red; manganese containing soils are purplish-black
    Soil profile to identify specific soil type

c. Identify and discuss two methods of soil conservation that would be effective in areas with a high number of insect pests for your crops. **2 points**

Methods such as no tillage, minimum tillage, and terracing do not work to reduce pests. The method has to both reduce pests and decrease erosion.

Shelterbelts	Trees decrease wind erosion and pest migration
Ally cropping	Trees decrease wind erosion and pest migration
Crop rotation	Roots always in ground; pests usually cannot feed on subsequent crop
Strip cropping	Plant roots decrease erosion; decreased pest migration
Sacrificial crop	Draws pests away from primary crop; roots remain in ground for erosion
Polyculture	Plant roots decrease erosion; decreased pest migration

d. Other than the soil erosion control methods that also work for pests, identify two other examples of integrated pest management that could be used to help control crop pests. **2 points**

Sterile males, predatory insects, pheromones, hormones, traps, repellants, Bt, small amounts of nonpersistent pesticides, and so on.

e. Explain the difference between weathering and erosion, giving a specific example of each. **4 points**

Weathering is the breakdown of rock into soil.
Examples: Mechanical—frost wedging; pressure release; thermal expansion and contraction; organisms such as worms, ants, or roots
Chemical—lichens and carbonic acid; sulfuric and nitric acid; oxidation, hydrolysis

Erosion is movement of soil from one place to another
Examples: Wind or water (sheet erosion, gully erosion, rill erosion)

2. a. Calculate the amount of gasoline a typical car would require to travel 500 miles. **1 point**

1 gallon/25 miles x 500 miles = 20 gallons

b. Calculate how much E85 a flex fuel car would require to travel the same distance. **1 point**

20 gallons x 1.5 = 30 gallons

c. Assume that 10 percent of U.S. cars were flex fuel and used 100 percent ethanol, calculate the decrease in U.S. gasoline consumption. Determine if the current supply of U.S. ethanol is sufficient to meet the needed difference.

140 billion gallons/year x .1 = 14.0 billion gallons/year **1 point**

Need 14 billion and subtract the current 4 billion = 10 billion gallons of ethanol short per year—so no, does not meet proposed demand **1 point**

d. Using ethanol lowers the consumption of gasoline. Describe two environmental costs of switching to ethanol and two environmental benefits of switching to ethanol. **4 points cost/benefit must be linked to consequence**

Benefit	Consequence
Decreased carbon dioxide emissions	Carbon dioxide is greenhouse gas; decreased global warming; decreased acid deposition
Decreased carbon monoxide emissions	Carbon monoxide prevents adequate oxygenation of human blood; decreased human health hazard
Decreased NO_x	Less acid deposition; less smog; improved visibility
Decreased particulates	Improved visibility; decreased human lung problems
Less drilling for oil	Less habitat disruption
Fewer oil spills during transport	Less habitat degradation; death of animals in spills
Less oil refining	Human respiratory problems decreased; decreased habitat disruption
Cost	Consequence
Increased intensive agriculture	Increased use of fertilizer and pesticides; increased erosion; increased irrigation
Potential conversion of food crops to alcohol crops	Human food needs not met
Distillation costs	Air pollution from carbon monoxide, VOCs (formaldehyde and acetic acid) and methanol; water pollution

e. Propose one policy that the United States could implement to convince consumers to purchase flex fuel vehicles and give one drawback to the policy. **2 points—one for policy and one for drawback**

Policies
tax rebates, subsidies, lower tag/registration/parking fees, mandated sales quotas

Drawbacks
E85 not readily available
need to buy a new vehicle
ethanol expensive currently
fill up more frequently because fewer miles to the gallon
Americans more comfortable with gasoline
education of the populace
compliance with new standards

3. a. Calculate the amount of feed required to get a hog to market weight. **2 points—1 for setup and 1 for answer**

250 lbs at slaughter-80 lbs feeder weight = 170 lbs to gain

170 lbs x 3 lbs corn/1 lb weight = 510 lbs corn

b. Calculate the number of humans that could be fed in a day on the corn to get one hog to market weight. **2 points**

510 lbs corn x 280 kcal/1 lb corn x 1 human/2,000 kcal = 71 humans

c. Calculate the total bushels of corn produced in a year in the United States. **1 point**

70×10^6 acres x 140 bushels/acre = 9.8×10^6 bushels per year

d. Describe two environmental problems associated with feedlots and intensive animal production, and explain their impact on the environment. **1 point for each problem; 1 point for explanation of impact for each problem**

Livestock waste	Solid waste disposal problem
Runoff	Surface water: increased BOD; decreased DO; increased turbidity; temperature changes; coliform contamination; increased nitrates and phosphates induce eutrophication. Groundwater: bacterial contamination; excess nitrates
Erosion; land compaction	Loss of soil fertility; loss of nutrients

e. Recently hogs have been selectively bred to have less fat and more muscle mass. Discuss one reason why the fat in pork was a human health concern. **1 point**

High-fat diets are linked to increased heart disease, increased arteriosclerosis, breast and prostate cancer, and diabetes.

4. a. Explain the difference between hazardous waste and municipal solid waste. Identify a specific example of each. **3 points**

Municipal solid waste is normal trash produced by homes and businesses. It would include such materials as food wastes, yard wastes, sewage sludge, paper, cardboard, glass, and textiles. Hazardous wastes would include discarded materials containing substances known to be toxic, mutagenic, carcinogenic, or teratogenic to living organisms. Also included in hazardous wastes are ignitable, corrosive, explosive, or highly reactive chemicals. Examples of hazardous wastes would be fluorescent light bulbs, fly ash from incinerators/power plants, heavy metals, pesticides, paints, or solvents.

b. Describe three features of a municipal waste landfill. 3 points

Fairly close to urban areas for transport
Geologically stable site
Impermeable liners to prevent leaching
Located away from groundwater
Pipes to remove leachate
Cover with soil
Large enough to allow use for a prolonged period

c. Discuss two environmental problems associated with incineration of municipal solid waste. 4 points—1 point for each problem and 1 point for the consequence of each problem

Problem	Consequence
Dioxins	Endocrine disruptor; teratogenic; immunotoxic; hepatotoxic; bioaccumulates
SO_2	Ecosystem acidification when combined with water vapor in atmosphere to form sulfuric acid
NO_x	Ecosystem acidification when combined with water vapor in atmosphere to form nitric acid; eutrophication due to increased nutrient availability
Mercury	Neurotoxin; dementia and neurological disorders
Carbon dioxide	Greenhouse gas that increases global warming; some ecosystem acidification when carbonic acid forms
Thermal pollution	Kills temperature sensitive species